HARMONIC MAPPINGS IN THE PLANE

Harmonic mappings in the plane are univalent complex-valued harmonic functions of a complex variable. Conformal mappings are a special case where the real and imaginary parts are conjugate harmonic functions, satisfying the Cauchy–Riemann equations. Harmonic mappings were studied classically by differential geometers because they provide isothermal (or conformal) parameters for minimal surfaces. More recently they have been actively investigated by complex analysts as generalizations of univalent analytic functions, or conformal mappings. Many classical results of geometric function theory extend to harmonic mappings, but basic questions remain unresolved.

This book is the first comprehensive account of the theory of planar harmonic mappings, treating both the generalizations of univalent analytic functions and the connections with minimal surfaces. Essentially self-contained, the book contains background material in complex analysis and a full development of the classical theory of minimal surfaces, including the Weierstrass–Enneper representation. It is designed to introduce nonspecialists to a beautiful area of complex analysis and geometry.

Peter Duren is Professor of Mathematics at the University of Michigan, Ann Arbor.

156　Harmonic Mappings in the Plane

HARMONIC MAPPINGS IN THE PLANE

PETER DUREN

University of Michigan

CAMBRIDGE
UNIVERSITY PRESS

PUBLISHED BY THE PRESS SYNDICATE OF THE UNIVERSITY OF CAMBRIDGE
The Pitt Building, Trumpington Street, Cambridge, United Kingdom

CAMBRIDGE UNIVERSITY PRESS
The Edinburgh Building, Cambridge CB2 2RU, UK
40 West 20th Street, New York, NY 10011-4211, USA
477 Williamstown Road, Port Melbourne, VIC 3207, Australia
Ruiz de Alarcón 13, 28014 Madrid, Spain
Dock House, The Waterfront, Cape Town 8001, South Africa

http://www.cambridge.org

First published 2004

Printed in the United States of America

Typeface Times Roman 10.25/13 pt.   *System* $\LaTeX 2_\varepsilon$ [TB]

A catalog record for this book is available from the British Library.

Library of Congress Cataloging in Publication Data
Duren, Peter L., 1935–
Harmonic mappings in the plane / Peter Duren.
p.   cm. – (Cambridge tracts in mathematics ; 156)
Includes bibliographical references and index.
ISBN 0-521-64121-7
1. Harmonic maps.   I. Title.   II. Series.
QA614.73.D87   2004
514′.74 – dc22   2003056516

ISBN 0 521 64121 7 hardback

Dedicated to the memory of
Glenn Schober
(1938–1991)

Contents

Contents

Preface

Harmonic mappings in the plane are univalent complex-valued harmonic functions whose real and imaginary parts are not necessarily conjugate. In other words, the Cauchy–Riemann equations need not be satisfied, so the functions need not be analytic. Although harmonic mappings are natural generalizations of conformal mappings, they were studied originally by differential geometers because of their natural role in parametrizing minimal surfaces. Only in the mid-1980s did harmonic mappings begin to attract widespread interest among complex analysts. The catalyst was a landmark paper by James Clunie and Terry Sheil-Small in 1984, pointing out that many of the classical results for conformal mappings have clear analogues for harmonic mappings. Since that time the subject has developed rapidly, although a number of basic problems remain unresolved. This book is an attempt to make this beautiful material accessible to a wider mathematical public.

Most of the book concerns harmonic mappings in the plane, but there are occasional excursions into higher dimensions, if only to provide counterexamples. As a general rule, the rich structure of theory in the plane does not extend to higher-dimensional space. In many instances, the properties of *analytic* univalent functions serve as models for generalizations to harmonic mappings, but other results are peculiar to analytic functions and do not extend to more general harmonic mappings. On the other hand, some results for harmonic mappings have no counterpart for conformal mappings. This is particularly true of the connections with minimal surfaces, which are developed in the final two chapters.

The book is dedicated to my collaborator and close friend Glenn Schober. I began writing it a few months before Glenn's untimely death in 1991 and had the benefit of discussing its contents with him as the project took shape. It would have been a better book if Glenn could have written it with me. In any event, it certainly reflects ideas and insights gained through our long association.

I am also grateful to Harold Shapiro, Walter Hengartner, and Terry Sheil-Small for teaching me essential things about harmonic mappings. Many people read and criticized early drafts of the book. First and foremost, Richard Laugesen went through much of the manuscript with a fine-toothed comb, spotted errors and ambiguities, and suggested many improvements. His generous help and constant encouragement were invaluable. Others who read and criticized portions of the manuscript were Walter Hengartner, Paul Greiner, John Pfaltzgraff, Željko Čučković, Michael Dorff, and Dmitry Khavinson. Their comments were helpful and are much appreciated. Paul Greiner also assisted in producing the figures drawn with *Mathematica*. Marcin Bownik was a great help in extracting the figures from the computer and preparing them for publication.

Finally, I would like to thank Lauren Cowles, David Tranah, and others at Cambridge University Press for expert advice and technical assistance, and for amazing patience with a long overdue manuscript.

Peter Duren
Ann Arbor, Michigan

1

Preliminaries

1.1. Harmonic Mappings

A real-valued function $u(x, y)$ is *harmonic* if it satisfies Laplace's equation:

$$\Delta u = \frac{\partial^2 u}{\partial x^2} + \frac{\partial^2 u}{\partial y^2} = 0.$$

A one-to-one mapping $u = u(x, y)$, $v = v(x, y)$ from a region D in the xy-plane to a region Ω in the uv-plane is a *harmonic mapping* if the two coordinate functions are harmonic. It is convenient to use the complex notation $z = x + iy$, $w = u + iv$ and to write $w = f(z) = u(z) + iv(z)$. Thus a complex-valued harmonic function is a harmonic mapping of a domain $D \subset \mathbb{C}$ if and only if it is *univalent* (or one-to-one) in D, that is, if $f(z_1) \neq f(z_2)$ for all points z_1 and z_2 in D with $z_1 \neq z_2$. Here $\mathbb{C}$ denotes the complex plane.

It must be emphasized that in this book the term "harmonic mapping" will always mean a *univalent* complex-valued harmonic function, except for occasional discussion of higher-dimensional analogues. Some writers use the term in a broader sense that does not require univalence.

A complex-valued function $f = u + iv$ is *analytic* in a domain $D \subset \mathbb{C}$ if it has a derivative $f'(z)$ at each point $z \in D$. The *Cauchy–Riemann equations*

$$\frac{\partial u}{\partial x} = \frac{\partial v}{\partial y}, \qquad \frac{\partial u}{\partial y} = -\frac{\partial v}{\partial x}$$

are an immediate consequence. Conversely, if f has continuous first partial derivatives and the Cauchy–Riemann equations hold, then f is analytic in D. (See Ahlfors [3] for information about analytic functions.) It follows from the Cauchy–Riemann equations (and from the existence of higher derivatives) that every analytic function is harmonic. A pair of functions (u, v) that satisfy the Cauchy–Riemann equations is said to be a *conjugate pair*, and v is called the *harmonic conjugate* of u. Hence, $-u$ is the harmonic conjugate of v. Strictly speaking, the conjugate function is determined locally only up to an additive constant. In a multiply connected domain the conjugate function need not be single-valued.

An analytic univalent function is called a *conformal mapping* because it preserves angles between curves. In fact, this angle-preserving property characterizes analytic functions among all functions with continuous first partial derivatives and nonvanishing Jacobians, because it implies that the Cauchy–Riemann equations are satisfied.

The object of this book is to study complex-valued harmonic univalent functions whose real and imaginary parts are not necessarily conjugate. As soon as analyticity is abandoned, serious obstacles arise. Analytic functions are preserved under composition, but harmonic functions are not. A harmonic function of an analytic function is harmonic, but an analytic function of a harmonic function need not be harmonic. The analytic functions form an algebra, but the harmonic functions do not. Even the square or the reciprocal of a harmonic function need not be harmonic. The inverse of a harmonic mapping need not be harmonic. The boundary behavior of harmonic mappings may be much more complicated than that of conformal mappings. It will be seen, nevertheless, that much of the classical theory of conformal mappings can be carried over in some way to harmonic mappings.

The *Jacobian* of a function $f = u + iv$ is

$$J_f(z) = \begin{vmatrix} u_x & v_x \\ u_y & v_y \end{vmatrix} = u_x v_y - u_y v_x,$$

where the subscripts indicate partial derivatives. If f is analytic, its Jacobian takes the form $J_f(z) = (u_x)^2 + (v_x)^2 = |f'(z)|^2$. For analytic functions f, it is a classical result that $J_f(z) \neq 0$ if and only if f is locally univalent at z. Hans Lewy showed in 1936 that this remains true for harmonic mappings. A relatively simple proof will be given in Chapter 2. In view of Lewy's theorem, harmonic mappings are either *sense-preserving* (or *orientation-preserving*) with $J_f(z) > 0$, or *sense-reversing* with $J_f(z) < 0$ throughout the domain D where f is univalent. If f is sense-preserving, then $\overline{f}$ is sense-reversing. Conformal mappings are sense-preserving.

The simplest examples of harmonic mappings that need not be conformal are the *affine mappings* $f(z) = \alpha z + \gamma + \beta \overline{z}$ with $|\alpha| \neq |\beta|$. Affine mappings with $\gamma = 0$ are *linear mappings*. It is important to observe that every composition of a harmonic mapping with an affine mapping is again a harmonic mapping: if f is harmonic, then so is $\alpha f + \gamma + \beta \overline{f}$.

Another important example is the function $f(z) = z + \frac{1}{2}\overline{z}^2$, which maps the open unit disk $\mathbb{D}$ onto the region inside a hypocycloid of three cusps inscribed in the circle $|w| = \frac{3}{2}$. To verify its univalence, suppose $f(z_1) = f(z_2)$

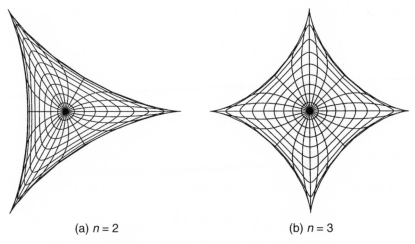

(a) $n = 2$ (b) $n = 3$

Figure 1.1. Image of mapping $f(z) = z + \frac{1}{n}\bar{z}^n$

for some points z_1 and z_2 in $\mathbb{D}$. Then

$$(\overline{z_1} + \overline{z_2})(\overline{z_1} - \overline{z_2}) = 2(z_2 - z_1).$$

But this is impossible unless $z_1 = z_2$, because $|z_1 + z_2| < 2$. The same argument shows that $f(z) = z + \frac{1}{n}\bar{z}^n$ is univalent for each $n \geq 2$.

The image of the disk under the mapping $f(z) = z + \frac{1}{n}\bar{z}^n$, as computed by *Mathematica*, is displayed graphically in Figure 1.1 for the cases $n = 2$ and 3. The curves in the figure are images of equally spaced concentric circles and radial segments. In general, the image of the disk under this mapping is bounded by a hypocycloid of $n + 1$ cusps inscribed in the circle $|w| = (n + 1)/n$.

In studying harmonic mappings of simply connected domains in the plane, there is no essential loss of generality in taking the unit disk as the domain of definition. To be more precise, suppose that f is a harmonic mapping of some simply connected domain $\Delta \subset \mathbb{C}$ onto a domain Ω, with $\Delta \neq \mathbb{C}$. The Riemann mapping theorem ensures the existence of a conformal mapping φ of $\mathbb{D}$ onto Δ. Thus the composition $F = f \circ \varphi$ is a harmonic mapping of $\mathbb{D}$ onto Ω. The original mapping is $f = F \circ \psi$, where ψ is the inverse of φ.

1.2. Some Basic Facts

Two simple differential operators appear commonly in complex analysis and are very convenient. They are

$$\frac{\partial}{\partial z} = \frac{1}{2}\left(\frac{\partial}{\partial x} - i\frac{\partial}{\partial y}\right) \quad \text{and} \quad \frac{\partial}{\partial \bar{z}} = \frac{1}{2}\left(\frac{\partial}{\partial x} + i\frac{\partial}{\partial y}\right),$$

where $z = x + iy$. For a complex-valued function $f(z)$, the equation $\partial f / \partial \overline{z} = 0$ is just another way of writing the Cauchy–Riemann equations. A direct calculation shows that the Laplacian of f is

$$\Delta f = 4 \frac{\partial^2 f}{\partial z \partial \overline{z}}.$$

Thus for functions f with continuous second partial derivatives, is is clear that f is harmonic if and only if $\partial f / \partial z$ is analytic. If f is analytic, then $\partial f / \partial z = f'(z)$, the ordinary derivative.

The operators $\partial / \partial z$ and $\partial / \partial \overline{z}$ are linear, and they have the usual properties of differential operators. For instance, the product and quotient rules hold:

$$\frac{\partial}{\partial z}(fg) = f \frac{\partial g}{\partial z} + g \frac{\partial f}{\partial z},$$

$$\frac{\partial}{\partial z}\left(\frac{f}{g}\right) = g^{-2}\left(g \frac{\partial f}{\partial z} - f \frac{\partial g}{\partial z}\right),$$

and similarly for $\partial / \partial \overline{z}$. The special property

$$\left(\frac{\partial f}{\partial z}\right)^{-} = \frac{\partial \overline{f}}{\partial \overline{z}}$$

connects the two derivatives. The differential

$$df = \frac{\partial f}{\partial x} dx + \frac{\partial f}{\partial y} dy$$

can be written as

$$df = \frac{\partial f}{\partial z} dz + \frac{\partial f}{\partial \overline{z}} d\overline{z},$$

thus motivating the notation $\partial / \partial z$ and $\partial / \partial \overline{z}$. The subscript notation $f_z = \partial f / \partial z$ and $f_{\overline{z}} = \partial f / \partial \overline{z}$ is often more convenient.

The chain rule for differentiation of composite functions can now be derived (formally). If $w = f(z)$ and $z = g(\zeta)$, then $w = h(\zeta)$, where $h = f \circ g$. Writing

$$dz = \frac{\partial g}{\partial \zeta} d\zeta + \frac{\partial g}{\partial \overline{\zeta}} d\overline{\zeta}$$

and

$$d\overline{z} = \frac{\partial \overline{g}}{\partial \zeta} d\zeta + \frac{\partial \overline{g}}{\partial \overline{\zeta}} d\overline{\zeta} = \overline{\frac{\partial g}{\partial \overline{\zeta}}} d\zeta + \overline{\frac{\partial g}{\partial \zeta}} d\overline{\zeta},$$

one finds after substitution that

$$dh = \frac{\partial f}{\partial z} \left(\frac{\partial g}{\partial \zeta} d\zeta + \frac{\partial g}{\partial \bar{\zeta}} d\bar{\zeta} \right) + \frac{\partial f}{\partial \bar{z}} \left(\frac{\overline{\partial g}}{\partial \bar{\zeta}} d\zeta + \frac{\overline{\partial g}}{\partial \zeta} d\bar{\zeta} \right).$$

Thus,

$$\frac{\partial h}{\partial \zeta} = \frac{\partial f}{\partial z} \frac{\partial g}{\partial \zeta} + \frac{\partial f}{\partial \bar{z}} \frac{\overline{\partial g}}{\partial \bar{\zeta}} \quad \text{and} \quad \frac{\partial h}{\partial \bar{\zeta}} = \frac{\partial f}{\partial z} \frac{\partial g}{\partial \bar{\zeta}} + \frac{\partial f}{\partial \bar{z}} \frac{\overline{\partial g}}{\partial \zeta}.$$

The Jacobian of a function $f = u + iv$ can be expressed as

$$J_f = |f_z|^2 - |f_{\bar{z}}|^2.$$

Consequently, f is locally univalent and sense-preserving wherever $|f_z(z)| > |f_{\bar{z}}(z)|$, and sense-reversing where $|f_z(z)| < |f_{\bar{z}}(z)|$. Note that $f_z(z) \neq 0$ wherever $J_f(z) > 0$. For sense-preserving mappings $w = f(z)$ one sees that

$$(|f_z| - |f_{\bar{z}}|)|dz| \leq |dw| \leq (|f_z| + |f_{\bar{z}}|)|dz|.$$

These sharp inequalities have the geometric interpretation that f maps an infinitesimal circle onto an infinitesimal ellipse with

$$D_f = \frac{|f_z| + |f_{\bar{z}}|}{|f_z| - |f_{\bar{z}}|}$$

as the ratio of the major and minor axes. The quantity $D_f = D_f(z)$ is called the *dilatation* of f at the point z. Clearly, $1 \leq D_f(z) < \infty$. A sense-preserving homeomorphism f is said to be *quasiconformal*, or *K-quasiconformal*, if $D_f(z) \leq K$ throughout the given region, where K is a constant and $1 \leq K < \infty$. The 1-quasiconformal mappings are simply the conformal mappings.

It is often more convenient to consider the ratio $\mu_f = f_{\bar{z}}/f_z$, called the *complex dilatation* of f. Thus, $0 \leq |\mu_f(z)| < 1$ if f is sense-preserving. It may be observed that $D_f(z) \leq K$ if and only if $|\mu_f(z)| \leq (K-1)/(K+1)$. It follows that a sense-preserving homeomorphism is quasiconformal if and only if its complex dilatation μ_f is bounded away from 1 in the given region: $|\mu_f(z)| \leq k < 1$. The mapping f is conformal if and only if $\mu_f = 0$. For the general theory of quasiconformal mappings the books by Lehto and Virtanen [1] and Ahlfors [1] are recommended.

In the theory of harmonic mappings, the quantity $\nu_f = \overline{f_{\bar{z}}}/f_z$, known as the *second complex dilatation*, turns out to be more relevant than the first complex dilatation μ_f. Since $|\nu_f| = |\mu_f|$, it is again clear that f is quasiconformal if and only if $|\nu_f(z)| \leq k < 1$.

Now let f be a complex-valued function defined in a domain $D \subset \mathbb{C}$ having continuous second partial derivatives. Suppose that f is locally univalent in D, with Jacobian $J_f(z) > 0$. Let $\omega = \nu_f = \overline{f_{\bar{z}}}/f_z$ be its second complex dilatation; then $|\omega(z)| < 1$ in D. Differentiating the equation $\overline{f_{\bar{z}}} = \omega f_z$ with respect to $\bar{z}$, one finds

$$\overline{f_{\bar{z}z}} = f_{z\bar{z}}\omega + f_z\omega_{\bar{z}}.$$

Now if f is harmonic in D, then $f_{z\bar{z}} = \frac{1}{4}\Delta f = 0$ there. Thus it follows that $\omega_{\bar{z}} = 0$ in D, so that ω is analytic. Conversely, if ω is analytic, then $\overline{f_{\bar{z}z}} = f_{z\bar{z}}\omega$. But since $|\omega(z)| < 1$, this implies that $f_{z\bar{z}} = 0$, and f is harmonic. Thus, f is harmonic if and only if ω is analytic. In particular, the second complex dilatation ω of a sense-preserving harmonic mapping f is always an analytic function of modulus less than one. This function ω will be called the *analytic dilatation* of f, or simply the dilatation when the context allows no confusion. Note that $\omega(z) \equiv 0$ if and only if f is analytic.

The analytic dilatation has some nice properties. For instance, if f is a sense-preserving harmonic mapping with analytic dilatation ω and it is followed by an affine mapping $A(w) = \alpha w + \gamma + \beta\overline{w}$ with $|\beta| < |\alpha|$, then the composition $F = A \circ f$ is a sense-preserving harmonic mapping with analytic dilatation

$$\frac{\overline{F_{\bar{z}}}}{F_z} = \frac{\overline{\alpha}\omega + \overline{\beta}}{\beta\omega + \alpha}.$$

For a proof, use the chain rule to calculate

$$F_z = A_w f_z + A_{\overline{w}}\overline{f_{\bar{z}}} = \alpha f_z + \beta\overline{f_{\bar{z}}},$$
$$F_{\bar{z}} = A_w f_{\bar{z}} + A_{\overline{w}}\overline{f_z} = \alpha f_{\bar{z}} + \beta\overline{f_z}.$$

Thus,

$$\frac{\overline{F_{\bar{z}}}}{F_z} = \frac{\overline{\alpha}\,\overline{f_{\bar{z}}} + \overline{\beta}\,f_z}{\beta\,\overline{f_{\bar{z}}} + \alpha f_z} = \frac{\overline{\alpha}\omega + \overline{\beta}}{\beta\omega + \alpha}.$$

The analytic dilatation also behaves well under precomposition. Let f be a sense-preserving harmonic mapping of a simply connected domain D onto a region Ω, with analytic dilatation ω. Let ψ map a domain Δ conformally onto D. Then the composition $F = f \circ \psi$ maps Δ harmonically onto Ω and has analytic dilatation $\omega \circ \psi$. To see this, simply use the chain rule to calculate $F_\zeta = f_z\psi'$ and $F_{\bar{\zeta}} = f_{\bar{z}}\overline{\psi'}$. Thus, the analytic dilatation of F is

$$\frac{\overline{F_{\bar{\zeta}}(\zeta)}}{F_\zeta(\zeta)} = \frac{\overline{f_{\bar{z}}(\psi(\zeta))}}{f_z(\psi(\zeta))} = \omega(\psi(\zeta)).$$

In a similar way, the first complex dilatation $\mu = f_{\bar{z}}/f_z$ shows a true invariance property. If f is followed by a conformal mapping φ and $F = \varphi \circ f$, then

F has the same complex dilatation μ. Indeed, the chain rule gives $F_z = \varphi' f_z$ and $F_{\bar{z}} = \varphi' f_{\bar{z}}$, so that $F_{\bar{z}}/F_z = f_{\bar{z}}/f_z$.

In a simply connected domain $D \subset \mathbb{C}$, a complex-valued harmonic function f has the representation $f = h + \overline{g}$, where h and g are analytic in D; this representation is unique up to an additive constant. For a proof, recall that f_z is analytic if f is harmonic, and let $h' = f_z$, where h is analytic in D. Now let $g = \overline{f} - \overline{h}$ and observe that

$$g_{\bar{z}} = \overline{f_z} - \overline{h_z} = 0 \qquad \text{in} \quad D$$

by the definition of h. Thus, g is analytic in D. The uniqueness of the representation depends on the fact that a function both analytic and anti-analytic must be constant. (An *anti-analytic function* is defined as the conjugate of an analytic function.) If f is real-valued, the representation reduces to $f = h + \overline{h} = \text{Re}\{2h\}$, where $2h$ is the analytic completion of f, unique up to an additive imaginary constant. In a multiply connected domain, the representation $f = h + \overline{g}$ is valid locally but may not have a single-valued global extension.

For a harmonic mapping f of the unit disk $\mathbb{D}$, it is convenient to choose the additive constant so that $g(0) = 0$. The representation $f = h + \overline{g}$ is then unique and is called the *canonical representation* of f.

1.3. The Argument Principle

First recall the classical argument principle for analytic functions and its elegant proof. Let D be a domain bounded by a rectifiable Jordan curve C, oriented in the positive or "counterclockwise" direction. Let f be analytic in D and continuous in $\overline{D}$, with $f(z) \neq 0$ on C. The *index* or *winding number* of the image curve $f(C)$ about the origin is $I = (1/2\pi)\Delta_C \arg f(z)$, the total change in the argument of $f(z)$ as z runs once around C, divided by 2π. Let N be the total number of zeros of f in D, counted according to multiplicity. The argument principle asserts that $N = I$.

The customary proof begins with the observation that f'/f has a simple pole with residue n wherever f has a zero of order n, so the residue theorem gives

$$N = \frac{1}{2\pi i} \int_C \frac{f'(z)}{f(z)}\, dz = \frac{1}{2\pi i} \Delta_C \log f(z) = I.$$

(Actually, since the derivative $f'(z)$ need not be defined on C, the curve of integration should be slightly contracted.) As an application, it can be seen that if f is analytic in D and continuous in $\overline{D}$, and if it carries C in a sense-preserving manner onto a Jordan curve Γ bounding a domain Ω, then f maps D

univalently onto Ω. In other words, univalence on the boundary implies univalence in the interior.

Because the argument principle has so many important applications, it will be very useful to have a generalization to complex-valued harmonic functions. In fact, the theorem is essentially of topological nature and may be generalized in various ways to arbitrary continuous mappings. However, it is desirable both to avoid the complications of topological degree theory and to develop a precise extension of the argument principle to "sense-preserving" harmonic functions. The proof for analytic functions suggests that the structure of harmonic functions may allow an elementary approach to a more general form of the theorem, and this turns out to be the case.

A complex-valued harmonic function f, not identically constant, will be classified as sense-preserving in a domain D if it satisfies a Beltrami equation of the second kind, $\overline{f_{\bar{z}}} = \omega f_z$, where ω is an analytic function in D with $|w(z)| < 1$. Since the Jacobian is $J_f = |f_z|^2 - |f_{\bar{z}}|^2$, this implies in particular that $J_f(z) > 0$ wherever $f_z(z) \neq 0$. If $f(z_0) = 0$ at some point z_0 in D, the order of the zero can be defined in terms of the canonical decomposition $f = h + \overline{g}$. Write the power-series expansions of h and g as

$$h(z) = a_0 + \sum_{k=n}^{\infty} a_k(z - z_0)^k, \qquad g(z) = b_0 + \sum_{k=m}^{\infty} b_k(z - z_0)^k,$$

where $n \geq 1$, $m \geq 1$, and $a_n \neq 0$, $b_m \neq 0$. (Here it is tacitly assumed that f is not analytic.) Actually, $b_0 = -\overline{a_0}$ because $f(z_0) = 0$. The sense-preserving property of f takes the equivalent form $g' = \omega h'$, with $|\omega(z)| < 1$. From this it follows that $m > n$, or that $m = n$ and $|b_n| < |a_n|$. In either case, we will say that f has a zero of *order n* at z_0.

As an immediate consequence of the structural formula, it can be inferred that the zeros of a sense-preserving harmonic function are isolated. Indeed, if $f(z_0) = 0$, then for $0 < |z - z_0| < \delta$ it is possible to write

$$f(z) = h(z) + \overline{g(z)} = a_n(z - z_0)^n \{1 + \psi(z)\},$$

where

$$\psi(z) = (\overline{b_m}/\overline{a_n})(\bar{z} - \overline{z_0})^m (z - z_0)^{-n} + \cdots.$$

But it is clear that $|\psi(z)| < 1$ for z sufficiently close to z_0, since $m \geq n$ and $|b_n/a_n| < 1$ if $m = n$. Hence $f(z) \neq 0$ elsewhere near z_0, and the zeros of f are isolated. Observe that the sense-preserving hypothesis is essential, because the zeros of a harmonic function are not always isolated. For example, the function $f(z) = z + \bar{z} = 2x$ vanishes at every point on the imaginary axis.

The argument principle for harmonic functions can now be formulated as a direct generalization of the classical result for analytic functions.

Theorem. *Let f be a sense-preserving harmonic function in a Jordan domain D with boundary C. Suppose f is continuous in $\overline{D}$ and $f(z) \neq 0$ on C. Then $\Delta_c \arg f(z) = 2\pi N$, where N is the total number of zeros of f in D, counted according to multiplicity.*

Proof. Suppose first that f has no zeros in D, so that $N = 0$ and the origin lies outside $f(D \cup C)$. A fact from topology says that in this case $\Delta_c \arg f(z) = 0$, which proves the theorem. To prove the topological fact, let ϕ be a homeomorphism of the closed unit square S onto $D \cup C$ with $\phi : \partial S \to C$ a homeomorphism. Then the composition $F = f \circ \phi$ is a continuous mapping of S onto the plane with no zeros, and we want to prove that $\Delta_{\partial S} \arg F(z) = 0$. Begin by subdividing S into finitely many small squares S_j on each of which the argument of $F(z)$ varies by at most $\pi/2$. Then $\Delta_{\partial S_j} \arg F(z) = 0$ and so

$$\Delta_{\partial S} \arg F(z) = \sum_j \Delta_{\partial S_j} \arg F(z) = 0,$$

where the first equality relies on the cancellation of contributions from the ∂S_j except on ∂S.

Next suppose that f does have zeros in D. Because the zeros are isolated and f does not vanish on C, there are only a finite number of distinct zeros in D. Denote them by z_j for $j = 1, 2, \ldots, \nu$. Let γ_j be a circle of radius $\delta > 0$ centered at z_j, where δ is chosen so small that the circles γ_j all lie in D and do not meet each other. Join each circle γ_j to C by a Jordan arc λ_j in D. Consider the closed path Γ formed by moving around C in the positive direction while making a detour along each λ_j to γ_j, running once around this circle in the negative (clockwise) direction, then returning along λ_j to C. This curve Γ contains no zeros of f, and so $\Delta_\Gamma \arg f(z) = 0$ by the case just considered. But the contributions of the arcs λ_j along Γ cancel out, so that

$$\Delta_C \arg f(z) = \sum_{j=1}^{\nu} \Delta_{\gamma_j} \arg f(z),$$

where each of the circles γ_j is now traversed in the positive direction. This formula reduces the global problem to a local one. (The same reduction is often used to prove the residue theorem.)

Suppose now that f has a zero of order n at a point z_0. Then, as observed earlier, f has the local form

$$f(z) = a_n(z - z_0)^n\{1 + \psi(z)\}, \qquad a_n \neq 0,$$

where $|\psi(z)| < 1$ on a sufficiently small circle γ defined by $|z - z_0| = \delta$. This shows that

$$\Delta_\gamma \arg f(z) = n \Delta_\gamma \arg \{z - z_0\} + \Delta_\gamma \arg \{1 + \psi(z)\} = 2\pi n.$$

Therefore, if f has zeros of order n_j at the points z_j, the conclusion is that

$$\Delta_C \arg f(z) = \sum_{j=1}^{\nu} \Delta_{\gamma_j} \arg f(z) = 2\pi \sum_{j=1}^{\nu} n_j = 2\pi N,$$

which proves the theorem. The result admits an obvious extension to multiply connected domains, just as for analytic functions. ∎

Several corollaries are worthy of note. First of all, there is a direct extension of Rouché's theorem to sense-preserving harmonic functions. Specifically, if p and $p + q$ are sense-preserving harmonic functions in D, continuous in $\overline{D}$, and $|q(z)| < |p(z)|$ on C, then p and $p + q$ have the same number of zeros inside D. As in the standard proof for analytic functions, the inequality on C implies that neither p nor $p + q$ has a zero on C and that the images of C under the two functions have the same winding numbers about the origin. Thus the harmonic version of Rouché's theorem follows from the harmonic version of the argument principle.

Next there is a generalization of Hurwitz's theorem. If f_n are harmonic functions in a domain D that converge locally uniformly, then their limit function f is harmonic. The harmonic version of Hurwitz's theorem asserts that if f and all of the f_n are sense-preserving, then a point z_0 in D is a zero of f if and only if it is a cluster point of zeros of the functions f_n. More precisely, f has a zero of order m at z_0 if and only if each small neighborhood of z_0 (small enough to contain no other zeros of f) contains precisely m zeros, counted according to multiplicity, of f_n for every n sufficiently large. The proof applies Rouché's theorem exactly as in the analytic case, with $p = f$ and $q = f_n - f$.

Finally, sense-preserving harmonic functions have the *open mapping property*: they carry open sets to open sets. In fact, as in the analytic case, a stronger statement can be made. If f is a sense-preserving harmonic function near a point z_0 where $f(z_0) = w_0$, and if $f(z) - w_0$ has a zero of order $n (n \geq 1)$ at z_0, then to each sufficiently small $\varepsilon > 0$ there corresponds a $\delta > 0$ with the following property. For each point $\alpha \in N_\delta(w_0) = \{w : |w - w_0| < \delta\}$, the function $f(z) - \alpha$ has exactly n zeros, counted according to multiplicity, in $N_\varepsilon(z_0)$. The proof appeals to the harmonic version of Rouché's theorem with $p = f - w_0$ and $q = w_0 - \alpha$.

The argument principle for harmonic functions has been essentially known for some time. Various forms of it have been applied in papers on harmonic mappings. However, the elementary proof presented here was found only recently by Duren, Hengartner, and Laugesen [1], who actually obtained a more general form of the theorem. As they pointed out, the proof still applies when $|\omega(z)| > 1$ in some parts of the domain D, so that f is sense-preserving in some regions and sense-reversing in others, provided that none of the zeros are situated at points where $|\omega(z)| = 1$. A zero at a sense-reversing point of f is assigned *negative* order, minus the order of the zero of $\overline{f}$ at the same point. Then a more general version of the theorem says that $\Delta_C \arg f(z)$ is equal to 2π times the sum of the orders of the zeros of f in D.

The classical version of the argument principle applies more generally to *meromorphic* functions and says that the winding number is equal to the number of zeros minus the number of poles, all counted according to multiplicity. Suffridge and Thompson [1] developed a form of the argument principle for harmonic functions that takes account of some kinds of singularities.

A less elementary proof of the argument principle proceeds through an important representation theorem for sense-preserving harmonic functions. This proof relies heavily on the theory of quasiconformal mappings and will only be sketched here. First contract the curve C to reduce the problem to the quasiconformal case where $|\omega(z)| \leq k < 1$ in D. Next observe that f satisfies a Beltrami equation of the *first* kind, $f_{\bar{z}} = \mu f_z$, where $\mu = (f_{\bar{z}}/\overline{f_z})\omega$. Appeal to standard results about quasiconformal mappings (see Lehto and Virtanen [1] or Ahlfors [1]) to conclude that f has the form $f = F \circ \Phi$, where Φ is a sense-preserving homeomorphism of $\overline{D}$ onto the closure of a Jordan domain Ω, and F is analytic in Ω. In this way the argument principle for sense-preserving harmonic functions is reduced to the classical result for analytic functions.

1.4. The Dirichlet Problem

In this section some facts about harmonic functions are assembled for easy reference in later chapters. Proofs are omitted but can be found in textbooks on complex analysis, for instance by Ahlfors [3] or Nehari [2].

One corollary of Green's theorem in the plane is *Green's identity*:

$$\iint_D (u\Delta v - v\Delta u)\, dx\, dy = \int_C \left(u\frac{\partial v}{\partial n} - v\frac{\partial u}{\partial n} \right) ds,$$

where D is a Jordan domain with smooth boundary C, the real-valued functions u and v have continuous second partial derivatives in $\overline{D}$ (the closure of D), $\partial/\partial n$ denotes an outer normal derivative, and ds denotes an element of arclength. If u is harmonic in D, so that its Laplacian $\Delta u = 0$, then by choosing v to be constant we see u can have no net flux across the boundary: $\int_C (\partial u / \partial n) \, ds = 0$. From this it follows that every harmonic function has the *mean-value property*

$$u(z_0) = \frac{1}{2\pi} \int_0^{2\pi} u(z_0 + \rho e^{i\theta}) \, d\theta$$

for each point $z_0 \in D$ and all radii $\rho > 0$ sufficiently small. Conversely, a continuous function with the local mean-value property must be harmonic. From the mean-value property it is a short step to the *maximum principle*: a function harmonic in a domain D cannot have a local maximum or minimum at any point in D unless it is identically constant. Thus, if u is a nonconstant function harmonic in D and continuous in $\overline{D}$, it will attain its maximum and minimum values only on the boundary.

The *Dirichlet problem* is to find a function harmonic in a domain D and continuous in $\overline{D}$ that agrees with a prescribed continuous function on the boundary ∂D. The uniqueness of a solution is an immediate consequence of the maximum principle. Existence of a solution is more difficult to establish, but an elegant proof can be given with the help of subharmonic functions if the boundary is sufficiently nice (see Ahlfors [3], p. 245 ff.). In particular, a solution always exists if D is a Jordan domain. Much more generally, it can be shown that a solution always exists (for every prescribed continuous boundary function) if and only if the boundary of D has no degenerate components. A *degenerate* boundary component is a component consisting of a single point.

When the given domain is a disk, the Dirichlet problem can be solved explicitly. For simplicity, consider the unit disk $\mathbb{D} = \{z \in \mathbb{C} : |z| < 1\}$ and let φ be an arbitrary continuous function on the interval $[0, 2\pi]$ with $\varphi(0) = \varphi(2\pi)$. Then the *Poisson formula* is

$$u(re^{i\theta}) = \frac{1}{2\pi} \int_0^{2\pi} P(r, \theta - t) \varphi(t) \, dt, \qquad 0 \le r < 1,$$

where

$$P(r, t) = \frac{1 - r^2}{1 - 2r \cos t + r^2}$$

is the *Poisson kernel*. This function u is harmonic in $\mathbb{D}$ and continuous in $\overline{\mathbb{D}}$, and $u(e^{it}) = \varphi(t)$ on the unit circle $\mathbb{T} = \{z \in \mathbb{C} : |z| = 1\}$. Thus, u solves the Dirichlet problem for the unit disk.

Suppose now that the prescribed function φ is piecewise continuous but has a finite number of jump discontinuities, so that at certain points $\theta \in [0, 2\pi]$

the left- and right-hand limits $\varphi(\theta-)$ and $\varphi(\theta+)$ exist but $\varphi(\theta-) \neq \varphi(\theta+)$. Then the function $u(z)$ given by the Poisson integral is harmonic in $\mathbb{D}$ and has the radial limit

$$\lim_{r \to 1} u(re^{i\theta}) = \frac{1}{2}(\varphi(\theta-) + \varphi(\theta+)).$$

More generally, as the point z in the disk approaches the boundary point $e^{i\theta}$ along a linear segment at an angle $\alpha(0 < \alpha < \pi)$ with the tangent line, it can be shown that $u(z)$ tends to the corresponding weighted average

$$\frac{\alpha}{\pi}\varphi(\theta-) + \left(1 - \frac{\alpha}{\pi}\right)\varphi(\theta+).$$

The Poisson formula can be generalized to an arbitrary Jordan domain D with rectifiable boundary C with the help of *Green's function*. This is the function $G(z, \zeta)$, harmonic in $D\backslash\{\zeta\}$ for each point $\zeta \in D$, for which $G(z, \zeta) + \log|z - \zeta|$ is harmonic at ζ and $G(z, \zeta) = 0$ for $z \in C$. For any function $\varphi(z)$ continuous on C, the function

$$u(\zeta) = -\frac{1}{2\pi} \int_C \varphi(z) \frac{\partial G}{\partial n}(z, \zeta) \, ds, \qquad \zeta \in D,$$

is the solution to the Dirichlet problem. Green's function $G(z, \zeta)$ can be obtained from the solution to a special Dirichlet problem with boundary function $\log|z - \zeta|$.

If D is again a Jordan domain with boundary C, the *harmonic measure* of a closed arc $I \subset C$ is the function $u(z)$ harmonic in D and continuous in $\overline{D}$ except at the endpoints of I, with $u(z) = 1$ on the interior of I and $u(z) = 0$ on $C\backslash I$. For example, if D is the unit disk and I is an arc of the unit circle with endpoints $e^{i\sigma}$ and $e^{i\tau}$, subtending an angle $\theta = \tau - \sigma$ $(0 < \theta < 2\pi)$ at the center of the disk, the harmonic measure of I has the form $u(z) = (1/\pi)(\alpha - (\theta/2))$, where $\alpha = \alpha(z)$ is the angle that I subtends at z.

With the help of the Poisson formula, it is easy to derive *Harnack's inequality*,

$$\frac{R-r}{R+r}u(0) \leq u(z) \leq \frac{R+r}{R-r}u(0), \qquad r = |z|,$$

for a *positive* harmonic function $u(z)$ in the disk $|z| < R$.

1.5. Conformal Mappings

Much of the theory of harmonic mappings is inspired by the classical theory of conformal mappings, a very special case. For later reference, we give here a rapid survey of conformal mappings and related topics. Proofs of theorems and further information can be found in the books by Nehari [2], Ahlfors [3], Pommerenke [1], and Duren [2].

A *domain* is defined to be an open connected set. A domain is *simply connected* if its complement with respect to the extended complex plane $\widehat{\mathbb{C}}$ is connected. A *doubly connected* domain is one whose complement consists of two components.

A function f is said to be a *conformal mapping* of a domain $\Omega \subset \mathbb{C}$ onto a domain D if it is analytic in Ω and *globally univalent* (*i.e.*, one-to-one) and it maps Ω *onto* D, so that $f(\Omega) = D$. An analytic function f is said to be *locally univalent* in Ω if it is univalent in some neighborhood of each point in Ω. A necessary and sufficient condition for local univalence is that $f'(z) \neq 0$ in Ω.

The famous *Riemann mapping theorem* asserts that every simply connected domain $\Omega \subset \mathbb{C}$ with $\Omega \neq \mathbb{C}$ admits a unique conformal mapping f onto the unit disk $\mathbb{D}$ with the properties $f(\zeta) = 0$ and $f'(\zeta) > 0$ for an arbitrarily prescribed point $\zeta \in \Omega$. Because the inverse function is necessarily analytic, it is equivalent to say that $\mathbb{D}$ can be mapped conformally onto Ω. The *Carathéodory extension theorem* says (in a special case) that each conformal mapping of a Jordan domain Ω onto a Jordan domain D can be extended to a homeomorphism of $\overline{\Omega}$ onto $\overline{D}$. This last theorem can be generalized to quasiconformal mappings.

The modern proof of Riemann's theorem is based on the theory of normal families. A collection $\mathscr{F}$ of functions f defined on a domain Ω is said to be a *normal family* if every sequence of functions in $\mathscr{F}$ has a subsequence that converges locally uniformly in Ω, meaning that it converges uniformly in some neighborhood of each point of Ω. In view of the Heine–Borel theorem, locally uniform convergence is the same as uniform convergence on each compact subset of Ω. One can show that $\mathscr{F}$ is a normal family if and only if each sequence of functions in $\mathscr{F}$ has a subsequence that converges uniformly in each compact subset of Ω. To see this, exhaust Ω by a sequence of expanding compacta and apply a diagonalization argument. A family $\mathscr{F}$ is said to be *locally bounded* in Ω if the functions in $\mathscr{F}$ are uniformly bounded in some neighborhood of each point of Ω. *Montel's theorem* says that a family of analytic functions is normal if and only if it is locally bounded. The proof essentially uses the Arzela–Ascoli theorem (*cf.* Rudin [1]) that a family is normal if it is equicontinuous and pointwise bounded. An application of the Cauchy integral formula shows that if a family $\mathscr{F}$ of analytic functions is locally bounded, then so is $\mathscr{F}' = \{f' : f \in \mathscr{F}\}$, and equicontinuity follows.

A standard result in complex analysis says that the locally uniform limit of a sequence of analytic functions is again analytic. The locally uniform limit of a sequence of analytic *univalent* functions is either univalent or constant.

The *Carathéodory convergence theorem* (see Duren [2], p. 76) relates the locally uniform convergence of a sequence of univalent analytic functions to a notion of convergence of its sequence of ranges. Let $\{D_n\}$ be a sequence of domains in the complex plane, each containing the origin. If the origin is an interior point of the intersection of the domains D_n, then the *kernel* of the sequence $\{D_n\}$ is defined as the largest domain D containing the origin and having the property that each compact subset of D lies in all but a finite number of the domains D_n. If the origin is not an interior point of the intersection, the kernel is defined as $D = \{0\}$. In either case, the sequence $\{D_n\}$ is said to *converge* to its kernel (written $D_n \to D$) if every subsequence has the same kernel. Now let f_n be a conformal mapping of the unit disk $\mathbb{D}$ onto a domain D_n, with $f_n(0) = 0$ and $f_n'(0) > 0$. Let D be the kernel of $\{D_n\}$. Then the Carathéodory convergence theorem says that $f_n \to f$ locally uniformly in $\mathbb{D}$ if and only if $D_n \to D \neq \mathbb{C}$. In the case of convergence, there are two possibilities. If $D = \{0\}$, then $f = 0$. If $D \neq \{0\}$, then D is a simply connected domain and f maps $\mathbb{D}$ conformally onto D.

The class S consists of all analytic univalent functions in $\mathbb{D}$, normalized so that $f(0) = 0$ and $f'(0) = 1$. Each function $f \in S$ has a power-series expansion of the form

$$f(z) = z + a_2 z^2 + a_3 z^3 + \cdots, \qquad |z| < 1.$$

The *Koebe function*

$$k(z) = \frac{z}{(1-z)^2} = z + 2z^2 + 3z^3 + \cdots$$

belongs to S and maps the disk onto the entire complex plane minus the portion of the negative real axis from $-\infty$ to $-\frac{1}{4}$. The *Koebe one-quarter theorem* says that the disk $|w| < \frac{1}{4}$ is contained in the range of every function in S. The Koebe function shows that the radius $\frac{1}{4}$ is best possible. *Bieberbach's theorem* asserts that $|a_2| \le 2$ for every function $f \in S$, with equality only for functions $f(z) = e^{-i\theta} k(e^{i\theta} z)$, rotations of the Koebe function. Bieberbach's theorem gives an easy proof of the Koebe one-quarter theorem and a wealth of other geometric information. It leads to the *Koebe distortion theorem*, which provides the sharp bounds

$$\frac{1-r}{(1+r)^3} \le |f'(z)| \le \frac{1+r}{(1-r)^3}, \qquad r = |z| < 1$$

for every $f \in S$. Again, equality occurs only for suitable rotations of the Koebe function. The Koebe distortion theorem leads in turn to the *growth theorem*

$$\frac{r}{(1+r)^2} \le |f(z)| \le \frac{r}{(1-r)^2}, \qquad r = |z| < 1$$

for $f \in S$, with equality only for suitable rotations of the Koebe function. Bieberbach's theorem can be generalized to say that the coefficients of every function $f \in S$ satisfy $|a_n| \leq n$ for all $n = 2, 3, \ldots$. This was known as the Bieberbach conjecture and was proved in 1984 by Louis de Branges. For a version of the proof along classical lines, see FitzGerald and Pommerenke [1].

1.6. Overview of Harmonic Mapping Theory

The development of a theory of harmonic mappings in the plane has come in two main stages. As early as the 1920s, differential geometers studied harmonic mappings because of their natural role in the theory of minimal surfaces. In any representation of a minimal surface by isothermal parameters, each of the three coordinate functions is harmonic. Thus, the projection of a nonparametric minimal surface onto its base plane induces a harmonic mapping. Properties of minimal surfaces such as Gauss curvature can be studied effectively through these harmonic mappings. Tibor Radó, Lipman Bers, Erhard Heinz, Johannes Nitsche, and others made early contributions (before 1960) to illuminate the interplay between harmonic mappings and minimal surfaces.

More recently, complex analysts have been interested in harmonic mappings as generalizations of conformal mappings. In a seminal paper, James Clunie and Terry Sheil-Small found viable analogues of the classical growth and distortion theorems, covering theorems, and coefficient estimates in this more general setting, although the sharp forms of the extended estimates are still largely undetermined. They also constructed a "harmonic Koebe function," which seems destined to play the extremal role of the Koebe function in the classical theory. That expectation gives rise to elegant and highly plausible conjectures, some of which have been verified for harmonic mappings with special geometric properties. The work of Clunie and Sheil-Small has attracted the attention of other complex analysts, and harmonic mappings have again become an active area of research.

Another aspect of the theory is the search for an appropriate "harmonic analogue" of the Riemann mapping theorem. Here the subject makes contact with partial differential equations and the well-developed theory of quasi-conformal mappings. In the 1980s, Walter Hengartner and Glenn Schober wrote a series of papers exploring this topic, among others. Standard results about quasiconformal mappings, specifically the existence of quasiconformal homeomorphisms as solutions of a Beltrami equation, suggested the following broad generalization of Riemann's theorem. Given a simply connected domain Ω and a point w_0 in Ω, and given an analytic function ω satisfying $|\omega(z)| < 1$ in the unit disk $\mathbb{D}$, there is a unique (sense-preserving) harmonic mapping f of $\mathbb{D}$ onto Ω with the prescribed analytic dilatation $\omega = \overline{f_{\bar{z}}}/f_z$ and with the properties $f(0) = w_0$ and $f_z(0) > 0$. However, the proposed extension of Riemann's theorem turned out to be false as stated. Hengartner and Schober found simple counterexamples and discovered a mysterious phenomenon of "collapsing" that typically prevents the existence of harmonic mappings with prescribed dilatation onto a given target region. They found, however, that the mapping does exist if the dilatation is further restricted or if the term "onto" is interpreted in a weaker sense. The question of uniqueness has not been fully settled, but the mapping is known to be unique if the target region is sufficiently nice.

The situation is relatively pleasant if the target region is convex. The harmonic mappings onto a given convex domain can be described completely in terms of boundary correspondences. This is the content of a remarkable result known as the Radó–Kneser–Choquet theorem. It shows that harmonic mappings are much more flexible than conformal mappings, which are completely determined by three consecutive boundary values. Among other applications, the theorem permits an effective study of extremal problems for convex harmonic mappings. Another consequence is the failure of Carathéodory's convergence theorem in the more general setting of harmonic mappings, but the manner of failure suggests how to formulate a correct generalization.

All of these topics will be developed in this book. The starting point will be an old theorem of Hans Lewy, already mentioned in Section 1.1, to the effect that local univalence of a complex-valued harmonic function is equivalent to the nonvanishing of its Jacobian. For analytic functions this phenomenon is well known.

Harmonic mappings in the plane have many remarkable properties that are not shared by their higher-dimensional analogues. Attempts to generalize these special results even to three-dimensional space seem doomed to failure, although some open questions of this sort still remain. This book will focus on planar harmonic mappings but will make occasional excursions into higher dimensions.

Survey articles by Schober [1] and Duren [3] may be consulted for further summary accounts of the subject up to about 1990.

2

General Properties of Harmonic Mappings

This chapter will develop a few basic properties of harmonic mappings, all generalizing well-known properties of conformal mappings. A preliminary discussion of critical points leads into a proof of Lewy's theorem that a locally univalent harmonic function has nonvanishing Jacobian. Then comes Radó's theorem that no harmonic function can map the disk univalently onto the whole plane.

2.1. Critical Points of Harmonic Functions

Let $u = u(x, y)$ be a real-valued function with continuous first partial derivatives in some region of the plane. A *critical point* of u is a point where $\partial u / \partial x$ and $\partial u / \partial y$ both vanish. Noncritical points are called *regular points*. It is obvious geometrically, or if one thinks in terms of mountain landscapes, that in general very little can be said about the structure of the critical set. It may contain isolated points (peaks), entire curves (horizontal ridges), and even open sets (plateaus). For a harmonic function, however, the critical set is always discrete.

Theorem. *All critical points of a nonconstant harmonic function are isolated.*

Proof. The critical points of u are precisely those where

$$\frac{\partial u}{\partial z} = \frac{1}{2} \left(\frac{\partial u}{\partial x} - i \frac{\partial u}{\partial y} \right) = 0.$$

But if u is harmonic, the function $\partial u / \partial z$ is analytic, and so its zeros are isolated unless $\partial u / \partial z \equiv 0$. ∎

It will be important to know the structure of the level-set of a harmonic function near a critical point, as described in the next theorem.

Theorem. *The level-set of a nonconstant harmonic function through a critical point z_0 consists locally of two or more analytic arcs intersecting with equal angles at z_0.*

Proof. Let $u = u(z)$ be harmonic in a neighborhood of the critical point z_0. Since the critical points of a harmonic function are isolated, we may choose a circular disk Δ about z_0 small enough to exclude all other critical points of u. Let $f = u + iv$ be an analytic completion of u in Δ. Since $\partial u / \partial x$ and $\partial u / \partial y$ both vanish at z_0, it follows from the Cauchy–Riemann equations that $\partial v / \partial x$ and $\partial v / \partial y$ also vanish there. Thus, $f'(z_0) = 0$.

Suppose for convenience that $f(z_0) = 0$ and that $z_0 = 0$. Then, near the origin, f has the form

$$f(z) = a_m z^m + a_{m+1} z^{m+1} + \cdots, \qquad a_m \neq 0,$$

for some integer $m \geq 2$. The set of points where $u(z) = \operatorname{Re}\{f(z)\} = 0$ is the preimage under f of the imaginary axis. Since $f(z)$ "behaves like a constant multiple of z^m near the origin," the stated property of the level-set of u is now intuitively clear. For a more rigorous treatment, write $f(z) = z^m \psi(z)$, where

$$\psi(z) = a_m + a_{m+1} z + a_{m+2} z^2 + \cdots \neq 0$$

in some neighborhood of the origin. Take a branch of the mth root to form the function

$$\varphi(z) = z[\psi(z)]^{1/m} = c_1 z + c_2 z^2 + \cdots, \qquad c_1 = a_m^{1/m},$$

which is analytic and univalent near the origin. Then f has the local structure $f(z) = [\varphi(z)]^m$, and $f(z)$ is purely imaginary exactly when the point $\varphi(z)$ lies on a system of m lines passing through the origin and meeting there with equal angles π/m. But φ is locally univalent and $\varphi(0) = 0$, so the preimage of that system of lines is locally a system of m analytic arcs passing through the origin and intersecting there with equal angles π/m. This is the local structure of the level-set of the harmonic function u, as the theorem asserts. ∎

If z_0 is a regular point of u, then $f'(z_0) \neq 0$ and f is locally univalent near z_0. It is then clear that the level-set of u is locally a single analytic arc passing through z_0. In particular, the level-set of a harmonic function cannot terminate at a regular point.

2.2. Lewy's Theorem

According to the inverse mapping theorem, a C^1 mapping from $\mathbb{R}^n$ to $\mathbb{R}^n$ with nonvanishing Jacobian is locally invertible (see, for instance, Rudin [1], p. 221). However, the nonvanishing of the Jacobian is in general not necessary for local invertibility. For example, the elemantary function $x \longmapsto x^3$ maps $\mathbb{R}^1$ univalently onto $\mathbb{R}^1$, yet its Jacobian vanishes at the origin. For *analytic* functions f, it is well known (see Ahlfors [3]) that the condition $f'(z_0) \neq 0$ is both necessary and sufficient for local univalence at z_0. Since the Jacobian is $J_f(z) = |f'(z)|^2$ for analytic functions, this says that the Jacobian of a locally univalent analytic function cannot vanish at any point. A theorem of Hans Lewy [1] asserts that the same principle holds more generally for harmonic functions in the plane.

Lewy's Theorem. *If f is a complex-valued harmonic function that is locally univalent in a domain $D \subset \mathbb{C}$, then its Jacobian $J_f(z)$ is different from 0 for all $z \in D$.*

Proof. Write $f = u + iv$ and suppose that $J_f(z_0) = 0$ for some point $z_0 \in D$. This means that the matrix

$$\begin{pmatrix} u_x & v_x \\ u_y & v_y \end{pmatrix}$$

has a vanishing determinant at z_0, so the homogeneous system of linear equations

$$au_x + bv_x = 0$$
$$au_y + bv_y = 0$$

has a nontrivial solution $(a, b) \neq (0, 0)$. In other words, the real-valued harmonic function $\psi = au + bv$ has a critical point at z_0. Suppose for convenience that $f(z_0) = 0$, and consider the level-set $\psi(z) = 0$ near the point z_0. As seen in the preceding section, this level-set consists locally of two or more distinct arcs that intersect with equal angles at z_0. On the other hand, f maps this level-set into the line $au + bv = 0$. But f is locally univalent at z_0, so it cannot carry a set composed of several intersecting arcs onto a linear segment. Thus, the assumption that $J_f(z_0) = 0$ has led to a contradiction. ∎

This proof, simpler than Lewy's original argument, appears in a 1951 paper by Lipman Bers [1]. However, the basic idea was used much earlier by Helmut Kneser [1] for a different purpose: to prove a theorem now known as the Radó–Kneser–Choquet theorem. A full discussion will be given in Section 3.1.

2.3. Heinz's Lemma

We now turn to a basic property of harmonic self-mappings of the disk. Those rather special mappings are treated in greater detail in Chapter 4, but it is convenient to establish a primitive version of the Heinz lemma at this point.

Heinz's Lemma. *Let f map the unit disk harmonically onto itself, with $f(0) = 0$. Then*

$$|f_z(0)|^2 + |f_{\bar{z}}(0)|^2 \geq c$$

for some absolute constant $c > 0$.

Erhard Heinz [1] discovered this lemma in 1952 and applied it to estimate the Gauss curvature of certain minimal surfaces. (Details are given in the last two chapters of this book; see especially Section 10.3.) His relatively simple proof led to the constant $c = 0.1788\dots$. Subsequently, J. C. C. Nitsche [2,4,6], H. L. de Vries [1,2], and Heinz [2] gave different proofs with improved values of the constant. They conjectured the sharp value $c = \frac{27}{4\pi^2} = 0.6839\dots$, and this was finally verified by R. R. Hall [2] in 1982. An account of Hall's proof appears later in this book (see Section 4.4). However, because the primitive form of the lemma already has important applications and is relatively easy to establish, a version of Heinz's original proof will now be presented.

One further remark may help to put Heinz's lemma in perspective. The quantity

$$|f_z|^2 + |f_{\bar{z}}|^2 = \frac{1}{2}\left(u_x^2 + u_y^2 + v_x^2 + v_y^2\right)$$

may be viewed as a measure of distortion under the mapping $f = u + iv$. Lewy's theorem says that the Jacobian $|f_z|^2 - |f_{\bar{z}}|^2$ is strictly positive in the disk, but even under the hypotheses of Heinz's lemma it has no absolute positive lower bound at the origin. (This will be seen later, at the end of Section 4.1, with the help of the Radó–Kneser–Choquet theorem.) Thus, the distortion theorem in its most natural form is false, and the Heinz lemma is a useful substitute.

Proof of Heinz's Lemma. The following approximation argument shows it is enough to assume that f extends to a homeomorphism of $\mathbb{D}$ onto itself. For $0 < r < 1$ let $D_r \subset \mathbb{D}$ be the preimage under f of the disk $|w| < r$, and let φ be a conformal mapping of $\mathbb{D}$ onto D_r with $\varphi(0) = 0$. Then $g = \frac{1}{r} f \circ \varphi$ maps

$\mathbb{D}$ harmonically onto itself with $g(0) = 0$ (since $f(0) = 0$), and g extends homeomorphically to the closure of $\mathbb{D}$. But

$$g_z(0) = \frac{1}{r} f_z(0) \varphi'(0) \qquad \text{and} \qquad g_{\bar{z}}(0) = \frac{1}{r} f_{\bar{z}}(0) \overline{\varphi'(0)},$$

so if it can be shown that $|g_z(0)|^2 + |g_{\bar{z}}(0)|^2 \geq c$, the desired inequality $|f_z(0)|^2 + |f_{\bar{z}}(0)|^2 \geq c$ will follow as r tends to one, because the Schwarz lemma gives $|\varphi'(0)| \leq 1$.

In view of the symmetry of the expression to be estimated, it may also be assumed without loss of generality that f is sense-preserving. Under these assumptions it is clear that f has a Poisson representation:

$$f(z) = \frac{1}{2\pi} \int_0^{2\pi} \frac{1 - |z|^2}{|e^{it} - z|^2} f(e^{it}) \, dt,$$

where $f(e^{it}) = e^{i\theta(t)}$ and $\theta(t)$ is continuous and strictly increasing on the interval $[0, 2\pi]$, with $\theta(2\pi) - \theta(0) = 2\pi$. In standard notation, f has the expansion

$$f(z) = \sum_{n=1}^{\infty} \left(a_n z^n + \bar{b}_n \bar{z}^n \right), \qquad a_1 = f_z(0), \quad b_1 = f_{\bar{z}}(0).$$

The Poisson kernel is

$$\frac{1 - |z|^2}{|e^{it} - z|^2} = \text{Re} \left\{ \frac{e^{it} + z}{e^{it} - z} \right\} = 1 + \sum_{n=1}^{\infty} (e^{-int} z^n + e^{int} \bar{z}^n).$$

Thus, the Poisson representation provides the formulas

$$a_n = \frac{1}{2\pi} \int_0^{2\pi} e^{-int} e^{i\theta(t)} \, dt, \qquad b_n = \frac{1}{2\pi} \int_0^{2\pi} e^{-int} e^{-i\theta(t)} \, dt.$$

Integration by parts now gives

$$2\pi n a_n = \int_0^{2\pi} e^{-int} e^{i\theta(t)} \, d\theta(t),$$

$$2\pi n b_n = - \int_0^{2\pi} e^{-int} e^{-i\theta(t)} \, d\theta(t).$$

On the other hand, it follows from Parseval's relation that

$$\sum_{n=1}^{\infty} \left(|a_n|^2 + |b_n|^2 \right) = \frac{1}{2\pi} \int_0^{2\pi} |f(e^{it})|^2 \, dt = 1.$$

Thus, it must be shown that

$$\sum_{n=2}^{\infty} \left(|a_n|^2 + |b_n|^2 \right) \leq 1 - c, \qquad c > 0.$$

The trivial estimates $|a_n| \leq \frac{1}{n}$ and $|b_n| \leq \frac{1}{n}$ are not good enough for this purpose. Instead, the formulas for a_n and b_n will be combined to obtain an improved estimate for $|a_n|^2 + |b_n|^2$. A short calculation leads to the expression

$$2\pi^2 n^2 \left(|a_n|^2 + |b_n|^2 \right) = \int_0^{2\pi} \int_0^{2\pi} \cos n(s - t) \cos\left[\theta(s) - \theta(t) \right] d\theta(s) \, d\theta(t)$$

$$\leq \int_0^{2\pi} d\theta(t) \int_0^{2\pi} |d_s \sin\left[\theta(s) - \theta(t) \right]|$$

$$= \int_0^{2\pi} d\theta(t) \int_0^{2\pi} |d \sin\theta| = 8\pi.$$

Applying this estimate, one finds

$$\sum_{n=2}^{\infty} \left(|a_n|^2 + |b_n|^2 \right) \leq \frac{4}{\pi} \sum_{n=2}^{\infty} \frac{1}{n^2} = \frac{2\pi}{3} - \frac{4}{\pi},$$

the desired inequality with $c = 1 - \frac{2\pi}{3} + \frac{4}{\pi} = 0.1778\ldots$. Heinz's inequality now follows from Parseval's relation. ∎

2.4. Radó's Theorem

It is well known that the only conformal mappings of the whole complex plane $\mathbb{C}$ onto itself are those of the form $f(z) = \alpha z + \beta$, where α and β are complex constants. One quick proof uses Picard's theorem to show that f cannot have an essential singularity at infinity; thus, f is a polynomial, which must be of first degree if it is univalent in the plane. The following theorem extends the result to harmonic mappings.

Theorem. *The only harmonic mappings of $\mathbb{C}$ onto $\mathbb{C}$ are the affine mappings $f(z) = \alpha z + \gamma + \beta \bar{z}$, where α, β, and γ are complex constants and $|\alpha| \neq |\beta|$.*

Proof. Let $f = h + \bar{g}$ map $\mathbb{C}$ harmonically onto $\mathbb{C}$, and assume without loss of generality that f is sense-preserving. This means that $|g'(z)| < |h'(z)|$, or $|g'(z)/h'(z)| < 1$ for all $z \in \mathbb{C}$. Hence, by Liouville's theorem, $g'(z)/h'(z) \equiv b$ for some complex constant b with $|b| < 1$. Integration gives

$$g(z) = bh(z) + c,$$

where c is a constant. Thus, f has the form

$$f = h + \bar{c} + \overline{bh} = F \circ h,$$

say, where F is an (invertible) affine mapping. It follows that $h = F^{-1} \circ f$ maps $\mathbb{C}$ univalently onto $\mathbb{C}$. But h is analytic, so it must have the form $h(z) = \alpha z + \beta$ for some complex constants α and β. This shows that f is an affine mapping. ∎

In fact, the proof shows that the only harmonic mappings of the plane *into* itself are the affine mappings, which actually send the plane *onto* itself. In other words, there exists no harmonic mapping of the plane onto a proper subdomain. In particular, no harmonic function can map $\mathbb{C}$ univalently onto the unit disk $\mathbb{D}$, a fact that also follows from Liouville's theorem.

In the opposite direction, it is easily seen that no *analytic* function can map $\mathbb{D}$ univalently onto $\mathbb{C}$. Indeed, the inverse of such a mapping would be analytic and bounded in $\mathbb{C}$ and, hence, constant. This argument does not apply to harmonic mappings, because the inverse need not be harmonic. Nevertheless, the result does extend to harmonic mappings.

Radó's Theorem. *There is no harmonic mapping of $\mathbb{D}$ onto $\mathbb{C}$.*

Proof. The following argument actually gives a stronger quantitative form of Radó's theorem. Suppose that f maps $\mathbb{D}$ harmonically onto a domain $\Omega \subset \mathbb{C}$ which contains a disk Δ_R of radius R. Assume without loss of generality that $f(0) = 0$ and $\Delta_R = \{w \in \mathbb{C} : |w| < R\}$. Denote by D_R the subdomain of $\mathbb{D}$ for which $f(D_R) = \Delta_R$. Let φ be a conformal mapping of $\mathbb{D}$ onto D_R with $\varphi(0) = 0$. Then $F = \frac{1}{R} f \circ \varphi$ maps $\mathbb{D}$ harmonically onto $\mathbb{D}$, with $F(0) = 0$, so the Heinz lemma says that

$$|F_\zeta(0)|^2 + |F_{\bar{\zeta}}(0)|^2 \geq c > 0,$$

where c is an absolute constant. But a calculation gives

$$F_\zeta(0) = \frac{1}{R} f_z(0)\varphi'(0); \qquad F_{\bar{\zeta}}(0) = \frac{1}{R} f_{\bar{z}}(0) \overline{\varphi'(0)}.$$

Since $|\varphi'(0)| \leq 1$ by the Schwarz lemma, it follows that

$$cR^2 \leq |f_z(0)|^2 + |f_{\bar{z}}(0)|^2.$$

In particular, the range of f cannot contain disks of arbitrarily large radius centered at the origin. ∎

Corollary. *No proper subdomain of the plane can be mapped harmonically onto the whole plane.*

Proof. Suppose there exists a harmonic mapping f of some simply connected domain $\Omega \neq \mathbb{C}$ onto $\mathbb{C}$. By the Riemann mapping theorem, there is a conformal mapping φ of $\mathbb{D}$ onto Ω. Thus, the composition $f \circ \varphi$ maps $\mathbb{D}$ harmonically onto $\mathbb{C}$, in violation of Radó's theorem. ∎

Tibor Radó [2] proved a special case of Radó's theorem in 1927. Proofs of the general theorem were given later by Bers [1] and Nitsche [2], both exploiting relations between harmonic mappings and minimal surfaces. Nitsche's approach exhibits a close connection with Heinz's lemma; in fact, he obtained an improved value of Heinz's constant as a corollary of his proof. The preceding derivation of Radó's theorem from Heinz's lemma was shown to the author by Harold Shapiro.

It is not known whether Radó's theorem extends to higher dimensions. In particular, Shapiro [1] has asked whether there exists a (univalent) harmonic mapping of the unit ball in $\mathbb{R}^3$ onto the whole space $\mathbb{R}^3$. Even this remains an open question.

2.5. Counterexamples in Higher Dimensions

The proof of Lewy's theorem in the plane, as presented in the previous section, exploits the local structure of the level-set of a harmonic function near a critical point. The proof would generalize to harmonic mappings in $\mathbb{R}^n$ if it could be shown that near a singular point the level-set of a harmonic function of n variables cannot be embedded into $\mathbb{R}^{n-1}$. (For instance, it might consist of two intersecting hypersurfaces.) However, Andrzej Szulkin [1] constructed a harmonic polynomial in $\mathbb{R}^3$ whose level-set near a critical point is a single surface, homeomorphic to a plane. His example is

$$\psi(x, y, z) = x^3 - 3xy^2 + z^3 - \frac{3}{2}(x^2 + y^2)z,$$

with a critical point at the origin. The reader is referred to Szulkin's paper for a proof that the zero-set of this homogeneous polynomial is a simple surface near the origin. This example shows that the foregoing *proof* of Lewy's theorem cannot work in $\mathbb{R}^n$ for any $n \geq 3$, but it does not disprove the theorem.

In fact, Lewy's theorem is false in dimensions higher than two. The following counterexample is due to J. C. Wood [1].

Consider the polynomial map f from $\mathbb{R}^3$ to $\mathbb{R}^3$ defined by $f(x, y, z) = (u, v, w)$, where

$$u = x^3 - 3xz^2 + yz, \qquad v = y - 3xz, \qquad w = z.$$

It is immediately verified that each component of f is a harmonic function in $\mathbb{R}^3$. To see that f is univalent, suppose that

$$f(x_1, y_1, z_1) = f(x_2, y_2, z_2) = (u, v, w)$$

for some pair of points (x_1, y_1, z_1) and (x_2, y_2, z_2) in $\mathbb{R}^3$. Then obviously $w = z_1 = z_2$ and

$$v = y_1 - 3x_1 w = y_2 - 3x_2 w,$$

which implies

$$u = x_1{}^3 + w(y_1 - 3x_1 w) = x_2{}^3 + w(y_2 - 3x_2 w).$$

It follows that

$$x_1{}^3 + vw = x_2{}^3 + vw,$$

so that $x_1 = x_2$ and $y_1 = y_2$. This proves the univalence of f. The calculations actually show that the mapping $(u, v, w) \longmapsto (x, y, z)$ defined by

$$x = (u - vw)^{1/3}, \qquad y = v + 3w(u - vw)^{1/3}, \qquad z = w,$$

is an inverse for f. Thus, f is a (univalent) harmonic mapping of $\mathbb{R}^3$ onto $\mathbb{R}^3$.

On the other hand, a straightforward calculation reveals that f has the Jacobian

$$J_f(x, y, z) = 3x^2,$$

which vanishes on the plane $x = 0$. Hence, Lewy's theorem is false in $\mathbb{R}^3$ and, therefore, in $\mathbb{R}^n$ for all $n \geq 3$.

Nevertheless, Lewy [2] was able to show that the theorem remains true in $\mathbb{R}^3$ under an additional hypothesis. A mapping f from $\mathbb{R}^n$ to $\mathbb{R}^n$ is called a *harmonic gradient mapping* if f is the gradient of a real-valued harmonic function $u(x_1, x_2, \ldots, x_n)$:

$$f(x_1, x_2, \ldots, x_n) = \left(\frac{\partial u}{\partial x_1}, \frac{\partial u}{\partial x_2}, \cdots, \frac{\partial u}{\partial x_n} \right).$$

For $n = 3$, Lewy proved that the Hessian of a harmonic function (the determinant of its matrix of second derivatives) cannot vanish at an interior point of its domain without changing sign, unless it vanishes identically. More precisely, if the Hessian vanishes at some interior point x_0 without vanishing identically, then in each neighborhood of x_0 it must take both positive and negative values. But the Jacobian of a harmonic mapping $f = \text{grad } u$ is the Hessian of u. As a consequence, the Jacobian of a locally univalent harmonic gradient mapping from $\mathbb{R}^3$ to $\mathbb{R}^3$ cannot vanish at any interior point of its domain. Gleason and Wolff [1] generalized this result to $\mathbb{R}^n$.

2.6. Approximation Theorem

The Carathéodory convergence theorem is a central result in the geometric theory of analytic functions, with important applications. It connects the *geometric* convergence of a sequence of simply connected domains with the *analytic* convergence of the corresponding sequence of Riemann mapping functions. (See Section 1.5 for a precise statement.) The Carathéodory theorem is not valid for harmonic mappings, a fact that often leads to complications. One immediate difficulty is the abundance of harmonic mappings with the same range. On the other hand, simple examples show that a sequence of harmonic mappings with common range may converge locally uniformly to a harmonic mapping with smaller range. An explicit construction is carried out in Section 4.1 on the basis of the Radó–Kneser–Choquet theorem.

Nevertheless, there is a special approximation theorem, due to Clunie and Sheil-Small [1], that sometimes acts as a substitute for one "half" of the Carathéodory convergence theorem. Before stating it, we need to introduce some terminology.

A function f_0 harmonic in the disk $\mathbb{D}$ is said to be *subordinate* to a harmonic function f if it has the form $f_0(z) = f(\omega(z))$ for some function ω analytic and univalent in $\mathbb{D}$ with the properties $|\omega(z)| < 1$ and $\omega(0) = 0$. If f is a harmonic mapping of $\mathbb{D}$ (a *univalent* harmonic function), and if Ω is a simply connected domain such that $f(0) \in \Omega \subset f(\mathbb{D})$, then there is a harmonic mapping f_0 of $\mathbb{D}$ onto Ω, subordinate to f. Furthermore, this mapping f_0 is unique up to rotation of the disk. To see this, we appeal to the Riemann mapping theorem and choose ω to be the conformal mapping of $\mathbb{D}$ onto $f^{-1}(\Omega)$ with $\omega(0) = 0$ and $\omega'(0) > 0$. The subordinate function $f_0 = f \circ \omega$ is then a harmonic mapping of $\mathbb{D}$ onto Ω, and it is uniquely determined by the normalization $\omega'(0) > 0$. Under these circumstances we will call f_0 *the* harmonic mapping subordinate to f, corresponding to the domain $\Omega \subset f(\mathbb{D})$.

As in the statement of the Carathéodory convergence theorem (*cf.* Section 1.5), a sequence of domains D_n is said to *converge* to a domain D if every subsequence of $\{D_n\}$ has D as its kernel.

Approximation Theorem. *Let f be a univalent harmonic function in $\mathbb{D}$, and let $\{\Omega_n\}$ be a sequence of simply connected domains with the property*

$f(0) \in \Omega_n \subset f(\mathbb{D})$ *that converges to* $f(\mathbb{D})$. *Then the corresponding sequence of subordinate functions* f_n *converges to* f *locally uniformly in* $\mathbb{D}$.

Proof. The subordinate functions have the form $f_n(z) = f(\omega_n(z))$, where ω_n are the analytic univalent functions defined earlier. The hypothesis that $\Omega_n \to f(\mathbb{D})$ is easily seen to imply that $\omega_n(\mathbb{D}) \to \mathbb{D}$. Thus, it follows from the Carathéodory convergence theorem that $\omega_n(z) \to z$ locally uniformly in $\mathbb{D}$. Thus, $f_n(z) \to f(z)$ locally uniformly.   ∎

In most applications of the approximation theorem, the domains Ω_n can be chosen to expand monotonically to their union $f(\mathbb{D})$. If Ω_n is taken to be a Jordan domain with smooth boundary, the corresponding function f_n will have a smooth extension to the closed disk.

3

Harmonic Mappings onto Convex Regions

This chapter deals with harmonic mappings of the unit disk onto convex regions. The simplest examples, the harmonic self-mappings of the disk, are singled out for detailed treatment in Chapter 4. The present chapter will focus on two important structural properties of convex mappings. The first is the celebrated Radó–Kneser–Choquet theorem, which constructs a harmonic mapping of the disk onto any bounded convex domain, with prescribed boundary correspondence. The second is the "shear construction" of a harmonic mapping with prescribed dilatation onto a domain convex in a given direction. This leads to an analytic description of convex mappings, which has various applications.

3.1. The Radó–Kneser–Choquet Theorem

Let $\Omega \subset \mathbb{C}$ be a domain bounded by a Jordan curve Γ. Each homeomorphism of the unit circle onto Γ has a unique harmonic extension to the unit disk $\mathbb{D}$, defined by the Poisson integral formula. The values of this harmonic extension must lie in the closed convex hull of Ω in view of the "averaging" property of the Poisson integral. It is a remarkable fact that if Ω is *convex*, this harmonic extension is always univalent and it maps the disk harmonically onto Ω.

This theorem was first stated in 1926 by Tibor Radó [1], who posed it as a problem in the *Jahresberichte*. Helmut Kneser [1] then supplied a brief but elegant proof. A period of almost 20 years elapsed before Gustave Choquet [1], apparently unaware of Kneser's note, rediscovered the result and gave a detailed proof that has some features in common with Kneser's but is not the same. In fact, the two approaches allow the theorem to be generalized in different directions. We shall present both proofs, beginning with Kneser's.

Radó–Kneser–Choquet Theorem. *Let $\Omega \subset \mathbb{C}$ be a bounded convex domain whose boundary is a Jordan curve Γ. Let φ be a homeomorphism of $\mathbb{T}$ onto Γ.*

Then its harmonic extension

$$f(z) = \frac{1}{2\pi} \int_0^{2\pi} \frac{1 - |z|^2}{|e^{it} - z|^2} \, \varphi\left(e^{it}\right) dt$$

is univalent in $\mathbb{D}$ and defines a harmonic mapping of $\mathbb{D}$ onto Ω.

Proof. Observe first that because the Poisson kernel is positive and has unit integral, the point $f(z)$ is a weighted average of points $\varphi(e^{it})$ distributed continuously around the boundary of the convex domain Ω, and so $f(z)$ must lie in Ω for every z in $\mathbb{D}$. The function f is harmonic in $\mathbb{D}$ and continuous in $\overline{\mathbb{D}}$, with boundary function

$$\hat{f}(\zeta) = \lim_{z \to \zeta} f(z) = \varphi(\zeta), \qquad \zeta \in \mathbb{T}.$$

(See Section 1.4 for properties of the Poisson integral.)

Without loss of generality, it may be supposed that the prescribed boundary function is sense-preserving: $\varphi(e^{it})$ runs around Γ in the positive (counter-clockwise) direction as t increases. (Otherwise, take complex conjugates.) The main difficulty of the proof is to establish the local univalence of f in $\mathbb{D}$ or, equivalently, the nonvanishing of its Jacobian. This will imply that f is sense-preserving in $\mathbb{D}$, in view of its approach to a sense-preserving boundary function. A direct application of the argument principle (Section 1.3) then confirms the global univalence and shows that f maps $\mathbb{D}$ onto Ω.

To prove the local univalence, suppose on the contrary that the Jacobian of f vanishes at some point z_0 in $\mathbb{D}$, so that the matrix

$$\begin{pmatrix} u_x & v_x \\ u_y & v_y \end{pmatrix}$$

has a vanishing determinant at z_0. As in the proof of Lewy's theorem (Section 2.2), it follows that the linear equations

$$au_x + bv_x = 0$$
$$au_y + bv_y = 0$$

have a nontrivial solution (a, b). Thus, the real-valued harmonic function $\psi = au + bv$ has a critical point at z_0. Let $c = \psi(z_0)$ and consider the level-set of all points z in $\mathbb{D}$ with $\psi(z) = c$. Denote by Λ the connected component of that level-set containing z_0. Observe that $f(\Lambda)$ lies in the intersection of Ω with the line $au + bv = c$. Near the critical point z_0, the set Λ consists of four or more arcs emanating from z_0 at equal angles (see Section 2.1). As these arcs extend away from z_0, they may conceivably branch as they meet other critical points. However, no arc of Λ can terminate in $\mathbb{D}$. (Again, see Section 2.1.)

Furthermore, no two of the arcs of Λ extending from z_0 can intersect either at another point of $\mathbb{D}$ or on the boundary $\mathbb{T}$. Indeed, such an intersection would entail the existence of a Jordan curve γ on which $\psi(z) = c$, and it would then follow from the maximum principle for harmonic functions that $\psi(z) = c$ everywhere inside γ. But this would imply, by the uniqueness principle for harmonic functions, that $\psi(z) = c$ throughout $\mathbb{D}$. In other words, it would imply that $f(\mathbb{D})$ lies on a line, which is clearly not the case. Therefore, no pair of the arcs of Λ emanating from z_0 can rejoin elsewhere in $\overline{\mathbb{D}}$, which means that Λ must meet $\mathbb{T}$ in at least four distinct points. On the other hand, f maps Λ into the line $au + bv = c$, which meets Γ in exactly two points because of the assumption that Ω is convex. Hence, f must map four or more points on $\mathbb{T}$ into at most two points on Γ, in violation of the hypothesis that f maps $\mathbb{T}$ in one-to-one fashion onto Γ. This contradiction proves that the Jacobian of f cannot vanish in $\mathbb{D}$, so f is locally univalent. The proof is now completed by appeal to the argument principle, as indicated. ∎

This geometric proof is due to Kneser [1]. Choquet's more analytic proof will be given in the next section. The theorem is clearly false if Ω is not convex, since then the range of f need not lie in Ω. However, as Kneser [1] remarked, his proof applies with no essential change to the case of a Jordan curve Γ bounding a nonconvex region Ω, provided it is assumed that $f(\mathbb{D}) \subset \Omega$. Then although the line $au + bv = c$ may meet Γ in many points, the (connected) image $f(\Lambda)$ is confined to the segment of that line that lies in $\overline{\Omega}$ and contains the point $w_0 = f(z_0)$. This segment meets Γ in exactly two points, so the argument goes through as before.

A more formal presentation of Kneser's proof may be found in a paper of Hildebrandt and Sauvigny ([1], pp. 78–80). Duren and Hengartner [2] and Lyzzaik [4] have adapted Kneser's basic idea to generalize the theorem to harmonic mappings of multiply connected domains. Here new complications arise because the level-curves of ψ may conceivably form loops surrounding inner boundary components, and it must be shown that this cannot happen. Harmonic mappings of multiply connected domains will be discussed later in this book, in Section 8.2.

3.2. Choquet's Proof

We now turn to Choquet's more analytic proof of the Radó–Kneser–Choquet theorem. He begins by observing, as had Kneser, that it is enough to establish the *local* univalence of f in $\mathbb{D}$. Still with Kneser, he argues that the vanishing of the Jacobian of $f = u + iv$ at some point z_0 in $\mathbb{D}$ would imply that some nondegenerate linear combination $\psi = au + bv$ has a critical point at z_0. Diverging from Kneser's proof (of which he was unaware), Choquet then appeals to the following lemma to reach a contradiction.

Lemma. *Let ψ be a real-valued function harmonic in $\mathbb{D}$ and continuous in $\overline{\mathbb{D}}$. If ψ is at most bivalent on $\mathbb{T}$, then ψ has no critical points in $\mathbb{D}$.*

To say that ψ is at most *bivalent* on $\mathbb{T}$ means that ψ takes any given value at most twice on $\mathbb{T}$. Let us first observe that the function $\psi = au + bv$ just constructed does have this property if Ω is strictly convex; *i.e.*, if the boundary Γ contains no line segments. Indeed, no line $au + bv = c$ can then intersect Γ in more than two points, and each of these points has a unique preimage under f, since f is assumed to map $\mathbb{T}$ in one-to-one fashion onto Γ. Thus, $\psi(z) = c$ for at most two points z on $\mathbb{T}$. The lemma then says that ψ can have no critical points in $\mathbb{D}$, which is a contradiction. In other words, the Radó–Kneser–Choquet theorem follows from the lemma under the extra assumption that Ω is strictly convex. As we shall see, the strict convexity is inessential and, in fact, the argument can be generalized to establish a much stronger form of the Radó–Kneser–Choquet theorem.

Proof of Lemma. It is to be shown that

$$\frac{\partial \psi}{\partial z} = \frac{1}{2}\left(\frac{\partial \psi}{\partial x} - i\frac{\partial \psi}{\partial y}\right) \neq 0$$

in $\mathbb{D}$. It will suffice to show that $\psi_z(0) \neq 0$, because this will imply that $\psi_z(z_0) \neq 0$ for each point z_0 in $\mathbb{D}$. To see this implication, let g be a conformal self-mapping of $\mathbb{D}$ with $g(0) = z_0$, and consider the composition $\Psi(\zeta) = \psi(g(\zeta))$. Observe that Ψ is again harmonic in $\mathbb{D}$, continuous in $\overline{\mathbb{D}}$, and at most bivalent on $\mathbb{T}$. A computation gives

$$\Psi_\zeta(\zeta) = \psi_z(g(\zeta))g'(\zeta),$$

since $g_{\bar{\zeta}}(\zeta) \equiv 0$. In particular, $\Psi_\zeta(0) = \psi_z(z_0)g'(0)$, so that $\Psi_\zeta(0) \neq 0$ implies that $\psi_z(z_0) \neq 0$.

To show that $\psi_z(0) \neq 0$, we use the Poisson representation

$$\psi(z) = \frac{1}{2\pi}\int_0^{2\pi} \mathrm{Re}\left\{\frac{e^{it} + z}{e^{it} - z}\right\}\psi(e^{it})\,dt$$

to calculate

$$\psi_z(z) = \frac{1}{2\pi} \int_0^{2\pi} \frac{e^{it}}{(e^{it} - z)^2} \, \psi(e^{it}) \, dt.$$

This gives the formula

$$\psi_z(0) = \frac{1}{2\pi} \int_0^{2\pi} e^{-it} \psi(e^{it}) \, dt.$$

On the other hand, the bivalence hypothesis says that $\psi(e^{it})$ can have only one local maximum and one local minimum on the circle, and that $\psi(e^{it})$ is monotonic on each of the arcs joining those points. After a rotation of coordinates, which does not change $|\psi_z(0)|$, we may conclude from the bivalence hypothesis that $\psi(e^{it})$ increases from a minimum at some point $e^{-i\alpha}$ to a maximum at $e^{i\alpha}$. Thus, ψ is strictly increasing as e^{it} moves in either direction from $e^{-i\alpha}$ to $e^{i\alpha}$, and so

$$\psi(e^{it}) - \psi(e^{-it}) > 0, \qquad 0 < t < \pi.$$

Consequently,

$$-2\pi \, \mathrm{Im} \, \{\psi_z(0)\} = \int_0^{2\pi} \psi(e^{it}) \sin t \, dt.$$

$$= \int_0^{\pi} [\psi(e^{it}) - \psi(e^{-it})] \sin t \, dt > 0,$$

proving that $\psi_z(0) \neq 0$.   ∎

Let us now observe that the lemma remains true under hypotheses on ψ much weaker than bivalence on $\mathbb{T}$. The essential requirement is that after a rotation of coordinates, the inequality $\psi(e^{it}) - \psi(e^{-it}) \geq 0$ should hold on the interval $[0, \pi]$, with strict inequality on some subinterval. This will happen whenever ψ is continuous on $\mathbb{T}$ and $\psi(e^{it})$ rises without decreasing from a minimum at $e^{-i\alpha}$ to a maximum at $e^{i\alpha}$, then falls without increasing to the minimum at $e^{-i\alpha}$ as the point e^{it} runs once around the circle. The conclusion is not affected if ψ remains constant on some arcs.

These considerations lead to a generalized version of the Radó–Kneser–Choquet theorem.

Radó–Kneser–Choquet Theorem (Strong Form). *Let $\Omega \subset \mathbb{C}$ be a bounded convex domain whose boundary is a Jordan curve Γ. Let φ map $\mathbb{T}$ continuously onto Γ and suppose that $\varphi(e^{it})$ runs once around Γ monotonically* *as e^{it} runs around $\mathbb{T}$. Then the harmonic extension*

$$f(z) = \frac{1}{2\pi} \int_0^{2\pi} \frac{1 - |z|^2}{|e^{it} - z|^2} \, \varphi(e^{it}) \, dt$$

is univalent in $\mathbb{D}$ and defines a harmonic mapping of $\mathbb{D}$ onto Ω.

It must be emphasized that in this theorem the prescribed boundary function φ is not required to be a homeomorphism; it may have arcs of constancy.

Proof of Theorem. As before, it must be shown that for each pair of real numbers $(a,b) \neq (0,0)$, the harmonic function $\psi = au + bv$ has the property $\psi_z(0) \neq 0$, where $f = u + iv$. The hypotheses ensure that the boundary function $\psi(e^{it})$ ascends monotonically from a minimum value to a strictly larger maximum value, then descends monotonically to its minimum value as e^{it} runs once around the circle. This is clear geometrically if one considers the family of parallel lines $au + bv = c$ for fixed (a,b). Thus, in light of the discussion following the proof of the lemma, it can be concluded that $\psi_z(0) \neq 0$, and the theorem follows. ∎

In fact, a modified form of the theorem remains true even if φ has points of discontinuity, provided that $\varphi(\mathbb{T})$ does not lie on a line and the values of φ go monotonically once around Γ. The harmonic extension f will then map $\mathbb{D}$ univalently onto the interior of the convex hull of $\varphi(\mathbb{T})$. For instance, if φ is piecewise constant and monotonic, and its values are not collinear, then f maps $\mathbb{D}$ univalently onto the interior of the convex polygon whose vertices are the values of φ. This function f has a rather curious boundary behavior: it effectively maps arcs to points and points to arcs. The general question of boundary behavior is discussed more fully in the next section.

3.3. Boundary Behavior

The Radó–Kneser–Choquet theorem has a partial converse that every harmonic mapping of the disk onto a strictly convex region can be extended continuously to the boundary. To say that a convex region is *strictly convex* means that its boundary contains no line segments. The boundary function need not be a homeomorphism; it can be constant on some arcs of the circle.

Choquet [1] stated this result on the boundary behavior of convex harmonic mappings for the special case where the range is a disk, attributing it to J. Deny (unpublished). Much later, in 1983, Choquet [2] wrote out a proof that was published in 1993. In fact, as Choquet [1] had also observed, the

result generalizes to harmonic mappings onto arbitrary Jordan domains, with
the proviso that a point of the circle may correspond to a linear segment on the
boundary of the range. Hengartner and Schober [5] gave an explicit version
of the generalized result, as follows.

Theorem. *Let f be a sense-preserving harmonic mapping of the unit disk $\mathbb{D}$
into a domain Ω bounded by a Jordan curve Γ in the finite plane, and suppose
that the radial limits $\lim_{r \to 1} f(re^{i\theta})$ lie on Γ for almost every θ. Then there
is a countable set $E \subset \mathbb{T}$ such that the unrestricted limit*

$$\hat{f}(e^{i\theta}) = \lim_{z \to e^{i\theta}} f(z)$$

exists at every point $e^{i\theta} \in \mathbb{T}\backslash E$ and lies on Γ. Furthermore,

(a) *$\hat{f}(e^{i\theta})$ is continuous and sense-preserving on $\mathbb{T}\backslash E$;*
(b) *the one-sided limits*

$$\hat{f}(e^{i\theta-}) = \lim_{t \nearrow \theta} \hat{f}(e^{it}) \quad and \quad \hat{f}(e^{i\theta+}) = \lim_{t \searrow \theta} \hat{f}(e^{it})$$

exist, belong to Γ, and are different for each point $e^{i\theta} \in E$;
(c) *the cluster set of f at each point $e^{i\theta} \in E$ is the linear segment joining
$\hat{f}(e^{i\theta-})$ to $\hat{f}(e^{i\theta+})$.*

It should be recalled that a bounded harmonic function has radial limits
almost everywhere. The continuity and sense-preserving properties of $\hat{f}$ in
(a) and the one-sided limits in (b) are to be taken relative to the set $\mathbb{T}\backslash E$,
since $\hat{f}$ is not defined on E. Of course, the exceptional set E may be empty,
in which case $\hat{f}$ is continuous on $\mathbb{T}$ and f has a continuous extension to $\overline{\mathbb{D}}$.
This will be the case if f maps $\mathbb{D}$ *onto* a strictly convex domain, because in
general the theorem asserts that each point of E must correspond to a linear
segment on the boundary of the range of f.

Proof of Theorem. Let φ be an analytic univalent function that maps $\mathbb{D}$ conformally onto Ω. By the Carathódory extension theorem, φ extends to a homeo-
morphism of $\overline{\mathbb{D}}$ onto $\overline{\Omega}$, with $\varphi(\mathbb{T}) = \Gamma$. The composition $g = \varphi^{-1} \circ f$ maps $\mathbb{D}$
into itself and has radial limits $\hat{g}(e^{i\theta})$ that lie on $\mathbb{T}$ for almost every θ. Since
g is sense-preserving, the boundary function has the same property, so it can
be extended to the full circle with the form

$$\hat{g}(e^{i\theta}) = e^{i\alpha(\theta)}, \qquad \theta \in \mathbb{R},$$

for some nondecreasing function $\alpha(\theta)$ with $\alpha(\theta + 2\pi) = \alpha(\theta) + 2\pi$. The extended boundary function $\hat{g}$ is continuous except perhaps for a countable set E where it has finite jumps. On $\mathbb{T} \backslash E$, the function $\varphi \circ \hat{g}$ is continuous and sense-preserving, with values on Γ. At each point of E, the one-sided limits of $\varphi \circ \hat{g}$ exist, lie on Γ, and are not equal. But $\varphi \circ \hat{g}$ agrees with the radial limit of f almost everywhere, so f can be recovered as a Poisson integral:

$$f(z) = \frac{1}{2\pi} \int_0^{2\pi} \frac{1 - |z|^2}{|e^{it} - z|^2} \varphi(e^{i\alpha(t)}) \, dt.$$

By properties of the Poisson integral (see Section 1.3), the unrestricted limit $\hat{f}(e^{i\theta})$ exists for each point $e^{i\theta} \in \mathbb{T} \backslash E$ and is equal to $\varphi(e^{i\alpha(\theta)})$. Thus, $\hat{f}$ inherits properties (a) and (b) from corresponding properties of $\varphi \circ \hat{g}$.

Finally, it is known that wherever a real-valued function has a jump point, its Poisson integral will cluster on the segment between the left- and right-hand limits and, in fact, will tend to a weighted average of these two values depending on the angle of approach to the boundary (see Section 1.4). For the Poisson integral of a complex-valued function this says that the cluster set at a jump discontinuity consists precisely of the linear segment joining the two one-sided limits.   ∎

It is clear that the proof extends to more general regions Ω, for example to bounded Jordan regions with finitely many internal slits. Hengartner and Schober [5] state the theorem for bounded simply connected domains with locally connected boundary, so that all prime ends degenerate to points and φ has a continuous extension to $\overline{\mathbb{D}}$.

3.4. The Shear Construction

The Radó–Kneser–Choquet theorem offers a feasible way to construct harmonic mappings of the disk onto any bounded convex region. The process will be applied and further explored in the next chapter. Meanwhile, it is useful to consider another general method for constructing harmonic mappings with specified properties. This method, introduced by Clunie and Sheil-Small [1], is known as the "shear construction." Essentially it produces a harmonic mapping onto a domain convex in one direction by "shearing" (or stretching and translating) a given conformal mapping along parallel lines. The dilatation of the resulting harmonic mapping can be prescribed, but only certain general features of its range are predetermined.

A domain $\Omega \subset \mathbb{C}$ is said to be *convex in the horizontal direction* (CHD) if its intersection with each horizontal line is connected (or empty). In other words, each line parallel to the real axis meets Ω in a full segment, possibly unbounded, or not at all. The basic theorem of Clunie and Sheil-Small is as follows.

Theorem 1. *Let $f = h + \bar{g}$ be harmonic and locally univalent in the unit disk. Then f is univalent and its range is CHD if and only if $h - g$ has the same properties.*

The proof relies on a simple lemma.

Lemma. *Let a domain $\Omega \subset \mathbb{C}$ be CHD, and let p be a real-valued continuous function on Ω. Then the mapping $w \longmapsto w + p(w)$ is univalent in Ω if and only if it is locally univalent. If it is univalent, then its range is CHD.*

Proof of Lemma. If the mapping is not univalent, then $w_1 + p(w_1) = w_2 + p(w_2)$ for some pair of points w_1 and w_2 in Ω, with $w_1 \neq w_2$. Write $w_1 = u_1 + iv_1$ and $w_2 = u_2 + iv_2$. Then $v_1 = v_2 = c$ (say), and the mapping $u \longmapsto u + p(u + ic)$ is not strictly monotonic and thus, is not locally univalent. In particular, the mapping $w \longmapsto w + p(w)$ cannot be locally univalent unless it is univalent in Ω. Geometrically, the mapping acts as a shear in the horizontal direction, so its range is clearly CHD. ∎

Proof of Theorem. Suppose first that $f = h + \bar{g}$ is univalent and that $\Omega = f(\mathbb{D})$ is CHD. Then the function

$$h(f^{-1}(w)) - g(f^{-1}(w)) = w - 2\operatorname{Re}\{g(f^{-1}(w))\} = w + p(w)$$

may be defined in Ω, where p is real-valued and continuous. But Lewy's theorem guarantees that $h'(z) \neq g'(z)$ in $\mathbb{D}$, so $h - g$ is locally univalent there. Hence, the mapping $w \longmapsto w + p(w)$ is locally univalent in Ω and so is univalent and has a range that is CHD, by the lemma. It follows that $h - g$ is univalent in $\mathbb{D}$ and its range is CHD.

Conversely, suppose that $F = h - g$ is univalent in $\mathbb{D}$ and that $\Omega = F(\mathbb{D})$ is CHD. Then

$$f(F^{-1}(w)) = w + 2\operatorname{Re}\{g(F^{-1}(w))\} = w + q(w)$$

is locally univalent in Ω and so (by the lemma) is univalent there and has a range that is CHD. Thus, f is univalent in $\mathbb{D}$ and its range is CHD. ∎

Observe now that a domain $\Omega \subset \mathbb{C}$ is convex if and only if it is convex in *every* direction. Thus a harmonic mapping $f = h + \bar{g}$ has convex range if and only if the range of every rotation $e^{i\alpha} f$ is CHD, for $0 \le \alpha < 2\pi$. In view of Theorem 1, it is equivalent to require that for each angle α the analytic function $e^{i\alpha} h - e^{-i\alpha} g$ be univalent and have a range that is CHD. The result can be summarized as follows.

Theorem 2. *Let $f = h + \overline{g}$ be harmonic and locally univalent in the unit disk. Then f is univalent and its range is convex if and only if for each choice of $\alpha (0 \le \alpha < 2\pi)$ the analytic function $e^{i\alpha} h - e^{-i\alpha} g$ is univalent and its range is CHD.*

Corollary. *If $f = h + \overline{g}$ is a convex harmonic mapping, then the function $h + e^{i\beta} g$ is univalent for each $\beta, 0 \le \beta < 2\pi$.*

Theorem 1 provides an effective device for constructing harmonic mappings with prescribed dilatation. Note that the prescription of a dilatation with $|\omega(z)| < 1$ immediately ensures the *local* univalence of the harmonic function f constructed from a given analytic univalent function. Indeed, the Jacobian of f is

$$J_f(z) = |h'(z)|^2 - |g'(z)|^2 = (1 - |\omega(z)|^2)|h'(z)|^2 > 0,$$

since

$$(1 - \omega(z))h'(z) = h'(z) - g'(z) \ne 0$$

by the univalence of $h - g$.

As a first example, let $h - g$ be the identity mapping and take as dilatation $\omega(z) = z$. Thus,

$$h(z) - g(z) = z \qquad \text{and} \qquad \frac{g'(z)}{h'(z)} = z.$$

Differentiation of the first equation gives the pair of linear equations

$$h'(z) - g'(z) = 1$$
$$zh'(z) - g'(z) = 0$$

with unique solution

$$h'(z) = \frac{1}{1 - z}, \qquad g'(z) = \frac{z}{1 - z}.$$

Integration now produces the expressions

$$h(z) = \log \frac{1}{1 - z}, \qquad g(z) = -z + \log \frac{1}{1 - z}$$

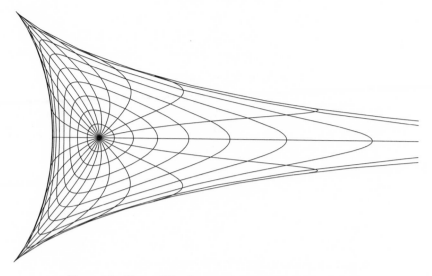

Figure 3.1. Shear of identity mapping with dilatation $\omega(z) = z$.

under the normalization $h(0) = g(0) = 0$. According to Theorem 1, the harmonic function $f = h + \overline{g}$, or

$$w = f(z) = -\overline{z} - 2\log|1 - z|,$$

maps the disk univalently onto a domain convex in the horizontal direction. In fact, the proof of Theorem 1 shows that the range of f is contained in the horizontal strip $|\text{Im}\{w\}| < 1$. Its actual range, as drawn by *Mathematica*, is shown in Figure 3.1, which also depicts the images of concentric circles and radial segments.

The example can be modified by prescribing the dilatation $\omega(z) = z^2$ instead of $\omega(z) = z$. The linear equations then become

$$h'(z) - g'(z) = 1$$
$$z^2 h'(z) - g'(z) = 0,$$

with normalized solution $h(z) = s(z)$ and $g(z) = -z + s(z)$, where

$$s(z) = \frac{1}{2} \log \frac{1+z}{1-z}$$

is a conformal mapping of the unit disk onto the horizontal strip $-\frac{\pi}{4} < \text{Im}\{w\} < \frac{\pi}{4}$. Thus, the shear construction produces a harmonic mapping

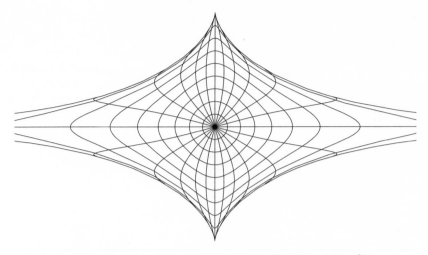

Figure 3.2. Shear of identity mapping with dilatation $\omega(z) = z^2$.

$f = h + \overline{g}$ of the form

$$f(z) = -\overline{z} + 2\operatorname{Re}\{s(z)\} = -\overline{z} + \log\left|\frac{1+z}{1-z}\right|.$$

Its image is depicted in Figure 3.2.

For a second example, consider the conformal mapping

$$w = \ell(z) = \frac{z}{1-z}$$

of the unit disk $\mathbb{D}$ onto the half-plane $\operatorname{Re}\{w\} > -\frac{1}{2}$. Theorem 1 can be restated to say that a locally univalent harmonic function $h + \overline{g}$ maps $\mathbb{D}$ univalently onto a region convex in the *vertical* direction if and only if $h + g$ has the same property. Thus, take $h + g = \ell$ and choose the dilatation $\omega(z) = -z$, which guarantees local univalence of $h + \overline{g}$. The resulting linear system

$$h'(z) + g'(z) = \ell'(z) = \frac{1}{(1-z)^2}$$
$$zh'(z) + g'(z) = 0$$

has the solution

$$h'(z) = \frac{1}{(1-z)^3}, \qquad g'(z) = -\frac{z}{(1-z)^3}.$$

Integration gives

$$h(z) = \frac{1}{2}[\ell(z) + k(z)], \qquad g(z) = \frac{1}{2}[\ell(z) - k(z)],$$

where $k(z) = z/(1-z)^2$ is the *Koebe function* (see Section 5.3), which maps $\mathbb{D}$ conformally onto the whole complex plane slit along the negative real axis from $-\frac{1}{4}$ to infinity. Thus, the harmonic function $L = h + \overline{g}$ maps $\mathbb{D}$ univalently onto a region convex in the vertical direction. Observe that L has the form

$$L(z) = \text{Re}\,\{\ell(z)\} + i\,\text{Im}\,\{k(z)\}.$$

We claim now that the range of L is the full half-plane $\text{Re}\,\{w\} > -\frac{1}{2}$. To see this, make the substitution $\zeta = \ell(z)$, so that $k(z) = \zeta(1+\zeta)$. Then, with the notation $\zeta = \xi + i\eta$, the harmonic mapping L takes the form

$$L(z) = \xi + i(1 + 2\xi)\eta, \qquad z = \ell^{-1}(\zeta) = \frac{\zeta}{1+\zeta}.$$

This shows that $L \circ \ell^{-1}$ maps each vertical line

$$\zeta = \xi_0 + i\eta, \qquad \xi_0 > -\frac{1}{2}, \qquad -\infty < \eta < \infty,$$

monotonically onto itself. These lines correspond to circles in the disk $|z| < 1$, internally tangent to the unit circle $\mathbb{T}$ at the point $z = 1$. In particular, the mapping $w = L(z)$ sends $\mathbb{D}$ univalently onto the half-plane $\text{Re}\,\{w\} > -\frac{1}{2}$.

The boundary correspondence under the harmonic mapping $L = \text{Re}\,\{\ell\} + i\,\text{Im}\,\{k\}$ is rather bizarre. Note that $\text{Re}\,\{\ell(z)\} = -\frac{1}{2}$ and $\text{Im}\,\{k(z)\} = 0$ for every point $z \neq 1$ on $\mathbb{T}$, since ℓ maps $\mathbb{D}$ conformally onto the half-plane $\text{Re}\,\{w\} > -\frac{1}{2}$ while k maps $\mathbb{D}$ onto the full plane minus part of the real axis. Consequently, $L(z) = -\frac{1}{2}$ for every point $z \neq 1$ on the unit circle! Figure 3.3 indicates this unusual behavior by showing the images under L of concentric circles and radial segments.

The example just constructed shows how radically the boundary behavior of harmonic mappings may differ from that of conformal mappings. According to the Carathéodory extension theorem, a conformal mapping between two Jordan domains always extends to a homeomorphism of the closures. In fact, Carathéodory's theorem generalizes to quasiconformal mappings (see Lehto and Virtanen [1], Ch. I, Sec. 8), which here amounts to requiring that the dilatation satisfy $|\omega(z)| \leq c < 1$ in $\mathbb{D}$.

For a third example, choose the conformal mapping

$$w = s(z) = \frac{1}{2}\log\frac{1+z}{1-z}$$

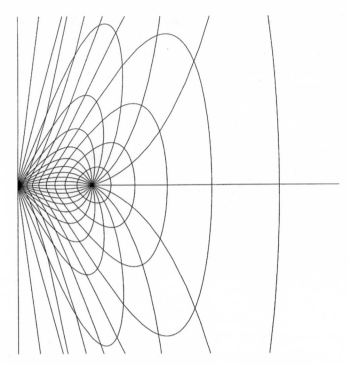

Figure 3.3. Vertical shear of half-plane mapping with dilatation $\omega(z) = -z$.

of $\mathbb{D}$ onto the horizontal strip $|\text{Im}\{w\}| < \frac{\pi}{4}$, and take first the dilatation $\omega(z) = z$. Then the relevant equations are

$$h(z) - g(z) = s(z) \qquad \text{and} \qquad zh'(z) - g'(z) = 0,$$

with normalized solution

$$h(z) = \frac{1}{2}(\ell(z) + s(z)), \qquad g(z) = \frac{1}{2}(\ell(z) - s(z)).$$

Thus, the harmonic mapping $f = h + \overline{g}$ is

$$f(z) = \text{Re}\{\ell(z)\} + i\,\text{Im}\{s(z)\}.$$

Observe that

$$f(e^{i\theta}) = \begin{cases} -\dfrac{1}{2} + i\dfrac{\pi}{4}, & 0 < \theta < \pi \\ -\dfrac{1}{2} - i\dfrac{\pi}{4}, & \pi < \theta < 2\pi. \end{cases}$$

In particular, f collapses the upper and lower semicircles to single points. In fact, it can be proved that f maps the disk precisely onto the half-strip

$$\left\{w : \text{Re}\{w\} > -\frac{1}{2}, \qquad |\text{Im}\{w\}| < \frac{\pi}{4}\right\}.$$

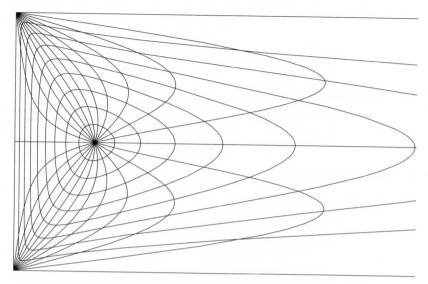

Figure 3.4. Shear of strip mapping with dilatation $\omega(z) = z$.

The image is shown in Figure 3.4. Note that all radial segments in the upper half-plane are mapped to arcs that terminate in the upper corner of the half-strip and similarly for the lower half-plane.

Finally, this last example can be modified by prescribing the dilatation $\omega(z) = z^2$. The resulting equations for h and g then have the normalized solutions $h = \frac{1}{2}(q + s)$ and $g = \frac{1}{2}(q - s)$, where

$$q(z) = \frac{z}{1 - z^2} = \sqrt{k(z^2)}$$

maps $\mathbb{D}$ conformally onto the whole plane minus the two radial slits from $\pm \frac{i}{2}$ to infinity. The corresponding harmonic mapping $f = h + \overline{g}$ is

$$f(z) = \operatorname{Re}\{q(z)\} + i \operatorname{Im}\{s(z)\},$$

which can be shown to map $\mathbb{D}$ onto the full strip $|\operatorname{Im}\{z\}| < \frac{\pi}{4}$. Observe that once again the upper and lower halves of the unit circle are sent to single points $\pm \frac{\pi}{4} i$, although those points are not now conspicuous geometric features of the range. The action of the mapping is shown in Figure 3.5.

Paul Greiner [1, 2] worked out further examples of the shear construction and applied *Mathematica* to display graphical images of concentric circles and radial lines in the style of the figures in this section. Driver and Duren [1] studied the harmonic shears of Schwarz–Christoffel mappings onto regular polygons, computing the integrals in terms of hypergeometric functions. Dorff and Szynal [1] computed the shears of complex elliptic integrals. Explicit

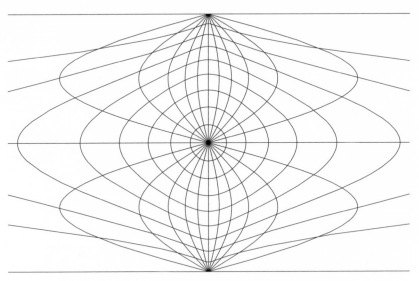

Figure 3.5. Shear of strip mapping with dilatation $\omega(z) = z^2$.

examples of the shear construction seem virtually unlimited; the present section gives only a small sample. In Section 5.3 the method will be applied, following Clunie and Sheil-Small, to construct the "harmonic Koebe function."

Finally, the idea of the shear construction leads to yet another proof of the Radó–Kneser–Choquet theorem. This approach will recur in Section 8.2, where it will be used to establish a generalized version of the theorem for multiply connected domains. Let $\Omega \subset \mathbb{C}$ be a bounded convex domain with boundary curve Γ. Let $w = \varphi(e^{it})$ be a sense-preserving homeomorphism of the unit circle $\mathbb{T}$ onto Γ, and let $f(z)$ be its harmonic extension to the unit disk $\mathbb{D}$. Let $f = h + \overline{g}$ be the canonical decomposition of f, and consider the analytic function

$$\phi_\alpha(z) = e^{i\alpha} h(z) - e^{-i\alpha} g(z) = e^{i\alpha} f(z) - 2 \operatorname{Re}\{e^{-i\alpha} g(z)\},$$

where α is a real parameter. Since f has convex range and ϕ_α is an open mapping, it is geometrically clear that the image of $\mathbb{T}$ under ϕ_α is a Jordan curve C_α bounding a domain that is convex in the direction of the real axis. Furthermore, as the point e^{it} runs once around $\mathbb{T}$, its image $\phi_\alpha(e^{it})$ runs once around C_α in the same direction. Thus the argument principle for analytic functions shows that ϕ_α is univalent in $\mathbb{D}$. In particular,

$$\phi_\alpha'(z) = e^{i\alpha} h'(z) - e^{-i\alpha} g'(z) \neq 0, \qquad z \in \mathbb{D}.$$

Since α is an arbitrary real number, it follows that

$$|h'(z)| \pm |g'(z)| \neq 0,$$

so that $|\omega(z)| \neq 1$ in $\mathbb{D}$. But f is sense-preserving near the boundary, so $|\omega(z)| < 1$ in $\mathbb{D}$. Now the argument principle for harmonic functions (see Section 1.3) takes over, as in the proofs of Kneser and Choquet, and allows the conclusion that f is globally univalent.

It must be observed that the proof just given is not quite complete because g was not shown to be continuous in $\overline{\mathbb{D}}$. One way around the difficulty is to replace $\mathbb{T}$ with a slightly smaller circle.

3.5. Structure of Convex Mappings

Conformal mappings of the unit disk onto convex domains have been studied for a long time and are known to have many special properties. They are described by the analytic condition

$$\mathrm{Re}\left\{1 + \frac{zf''(z)}{f'(z)}\right\} > 0, \qquad |z| < 1,$$

which essentially expresses the monotonic turning of the tangent vector at the boundary (see, for instance, Duren [2], Ch. 2). Implicit in this description is the *hereditary property*: if an analytic function maps the unit disk univalently onto a convex domain, then it also maps each concentric subdisk onto a convex domain.

It is natural to ask to what extent the special properties of conformal mappings will generalize to harmonic mappings of the disk onto convex domains. One obvious question is whether convexity remains a hereditary property under harmonic mappings. But the class of convex harmonic mappings $f = h + \overline{g}$ is neatly described by the theorem of Clunie and Sheil-Small as developed in the previous section (Section 3.4, Theorem 2); it says that f is univalent and convex if and only if for each $\alpha \in \mathbb{R}$ the analytic function $e^{i\alpha}h - e^{-i\alpha}g$ is univalent and convex in the horizontal direction. Thus, the hereditary question about convex harmonic mappings reduces to a similar question about analytic functions convex in one direction. Specifically, if an analytic function is univalent in the disk and its image is convex in the horizontal direction must the image of every concentric subdisk have the same property?

The answer is NO. Some years ago, Hengartner and Schober [2] showed that convexity in one direction is not a hereditary property for conformal mappings. Goodman and Saff [1] then constructed an example of a function convex in the vertical direction whose restriction to the disk $|z| < r$ does not have that property for any radius r in the interval $\sqrt{2} - 1 < r < 1$. They conjectured that the radius $\sqrt{2} - 1$ is best possible; in other words, each conformal mapping convex in a specified direction has that property when restricted to any disk of radius $r \leq \sqrt{2} - 1$. Ruscheweyh and Salinas [1] ultimately succeeded in proving the Goodman–Saff conjecture. Thus it follows, via the theorem of Clunie and Sheil-Small, that convex harmonic mappings have the corresponding property. To be more precise, if a function f maps the unit disk harmonically onto a convex domain, then for each radius $r \leq \sqrt{2} - 1$ it again maps the disk $|z| < r$ onto a convex domain, but it need not do so for any radius in the interval $\sqrt{2} - 1 < r < 1$.

There is a structural formula for analytic univalent functions convex in one direction, developed by Hengartner and Schober [1] in a special case, and later generalized by Royster and Ziegler [1] on the basis of earlier work by Robertson [1]. In principle, this representation should give complete information about convex harmonic functions. In reality, however, the formula is difficult to apply and other approaches are more effective. Ruscheweyh and Salinas used a method of convolution of power series (Hadamard products) in their proof of the Goodman–Saff conjecture. We will not pursue the details here.

Instead, we will show that the harmonic half-plane mapping

$$L(z) = \operatorname{Re}\left\{\ell(z)\right\} + i \operatorname{Im}\left\{k(z)\right\}$$

sends the subdisk $|z| < r$ onto a convex region for $r \leq \sqrt{2} - 1$, but onto a nonconvex region for $\sqrt{2} - 1 < r < 1$. Recall that this mapping L was constructed (see Section 3.4) by shearing the conformal mapping $\ell(z) = z/(1 - z)$ vertically with dilatation $\omega(z) = -z$. The function ℓ maps the unit disk conformally onto the half-plane $\operatorname{Re}\{w\} > -\frac{1}{2}$, and it often plays the role of extremal function in the family of convex conformal mappings, as does the Koebe function $k(z) = z/(1 - z)^2$ in the class of starlike mappings, or even in the full class of univalent functions. (In fact, $k(z) = z\ell'(z)$ and so ℓ and k are related by the canonical correspondence between convex and starlike mappings.) Thus, the harmonic mapping L, which maps the unit disk onto the same half-plane $\operatorname{Re}\{w\} > -\frac{1}{2}$, may be expected to play an extremal role in the class of convex harmonic mappings.

From this point of view it is not surprising that the function L exhibits the smallest "radius of convexity" in the family of convex harmonic mappings. Actually, the fact that convexity is not a hereditary property for harmonic mappings is already visually apparent from the graphical images obtained with

Mathematica and displayed in the last section; see in particular Figures 3.3 and 3.5. Indeed, a close inspection of Figure 3.3 reveals a breakdown of convexity of large level-curves of the mapping L near the point $w = -\frac{1}{2}$.

We shall now carry out a calculation to show that L maps the subdisks $|z| < r$ onto convex regions precisely for $r \leq \sqrt{2} - 1$ and not for any larger radius $r < 1$. For this purpose it will be necessary to study the change of the tangent direction

$$\Psi_r(\theta) = \arg\left\{\frac{\partial}{\partial\theta}L(re^{i\theta})\right\}$$

of the image curve as the point $z = re^{i\theta}$ moves around the circle $|z| = r$. Note first that

$$\frac{\partial}{\partial\theta}L(z) = \operatorname{Re}\left\{\frac{\partial}{\partial\theta}\ell(z)\right\} + i\operatorname{Im}\left\{\frac{\partial}{\partial\theta}k(z)\right\},$$

where

$$\frac{\partial}{\partial\theta}\ell(z) = iz\ell'(z) = \frac{iz}{(1-z)^2},$$

$$\frac{\partial}{\partial\theta}k(z) = izk'(z) = \frac{iz(1+z)}{(1-z)^3}.$$

Hence, a direct calculation gives

$$\frac{\partial}{\partial\theta}L(z) = A(r,\theta) + iB(r,\theta),$$

where

$$|1-z|^4 A(r,\theta) = r(r^2-1)\sin\theta$$

and

$$|1-z|^6 B(r,\theta) = r(1-r^4)\cos\theta - 2r^2(1-r^2)(1+\sin^2\theta).$$

The problem now reduces to finding the values of r such that the argument of the tangent vector, or equivalently $\tan\Psi_r(\theta)$, is a nondecreasing function of θ for $0 < \theta < \pi$. (Note that the curve $w = L(re^{i\theta})$ is symmetric with respect to the real axis, so it is sufficient to consider the interval $0 < \theta < \pi$.) But the formulas for $A(r,\theta)$ and $B(r,\theta)$ give

$$\tan\Psi_r(\theta) = \frac{B(r,\theta)}{A(r,\theta)} = \frac{2r(\csc\theta + \sin\theta) - (1+r^2)\cot\theta}{1 - 2r\cos\theta + r^2},$$

and lengthy calculation leads to an expression for the derivative in the form

$$(1 - u^2)|1 - z|^4 \frac{\partial}{\partial \theta} \tan \Psi_r(\theta) = p(r, u),$$

where $u = \cos \theta$ and

$$p(r, u) = 1 - 6r^2 + r^4 + 12r^2u^2 - 4r(1 + r^2)u^3.$$

The problem is now to find the values of the parameter r for which the cubic polynomial $p(r, u)$ is nonnegative in the whole interval $-1 \leq u \leq 1$. Observe first that

$$p(r, -1) = (1 + r)^4 > 0; \qquad p(r, 1) = (1 - r)^4 > 0.$$

Differentiation gives

$$\frac{\partial}{\partial u} p(r, u) = 12ru \left\{ 2r - (1 + r^2)u \right\},$$

showing that $p(r, u)$ has a local minimum at $u = 0$ and a local maximum at $u = 2r/(1 + r^3)$. The conclusion is that $p(r, u) \geq 0$ for $-1 \leq u \leq 1$ if and only if

$$p(r, 0) = 1 - 6r^2 + r^4 \geq 0.$$

But $1 - 6r^2 + r^4 \geq 0$ precisely for $r^2 \leq 3 - 2\sqrt{2}$, or $r \leq \sqrt{2} - 1$. This proves that the tangent angle $\Psi_r(\theta)$ increases monotonically with θ if $r < \sqrt{2} - 1$ but is not monotonic for $\sqrt{2} - 1 < r < 1$. Thus, the harmonic mapping L sends each disk $|z| < r \leq \sqrt{2} - 1$ to a convex region, but the image is not convex when $\sqrt{2} - 1 < r < 1$.

3.6. Covering Theorems and Coefficient Bounds

The classical Koebe one-quarter theorem says that each function $f(z) = z + a_2z^2 + \cdots$ analytic and univalent in the unit disk $\mathbb{D}$ contains the entire disk $|w| < \frac{1}{4}$ in its range $f(\mathbb{D})$. The Koebe function

$$k(z) = \frac{z}{(1 - z)^2} = z + 2z^2 + 3z^3 + \cdots$$

maps $\mathbb{D}$ conformally onto the full plane minus the portion of the negative real axis from $-\frac{1}{4}$ to infinity, showing that the radius $\frac{1}{4}$ is the best possible. The celebrated Bieberbach conjecture, now a theorem, asserts that the coefficients of each such function f satisfy the sharp inequalities $|a_n| \leq n, n = 2, 3, \ldots$.

Both of these results can be improved under the additional assumption that the range of f is convex. Let C denote the class of functions

$f(z) = z + a_2z^2 + \cdots$ that map the unit disk conformally onto a convex region. It is known that the range of each function $f \in C$ contains the larger disk $|w| < \frac{1}{2}$, and its coefficients satisfy the better bound $|a_n| \leq 1$. The function

$$\ell(z) = \frac{z}{1-z} = z + z^2 + z^3 + \cdots,$$

which maps $\mathbb{D}$ conformally onto the half-plane Re $\{w\} > -\frac{1}{2}$, shows that both results are again best possible (see Duren [2], p. 45).

These last results on convex conformal mappings extend nicely to convex harmonic mappings. Before stating the theorems, we need to introduce some terminology. The class C_H consists of all sense-preserving harmonic mappings $f = h + \overline{g}$ of the unit disk onto convex domains, with the normalization $h(0) = g(0) = 0$ and $h'(0) = 1$. Note that $|g'(0)| < |h'(0)| = 1$, since f preserves orientation and thus has positive Jacobian. Postcomposing a function $f = h + \overline{g} \in C_H$ by the sense-preserving affine mapping

$$\varphi(w) = \frac{w - \overline{b_1 \overline{w}}}{1 - |b_1|^2}, \qquad b_1 = g'(0),$$

which preserves convexity, the further normalization $g'(0) = 0$ can be achieved. The resulting class of functions will be denoted by C_H^0. Thus, a sense-preserving harmonic function $f = h + \overline{g}$ belongs to C_H^0 if it maps the unit disk univalently onto a convex region and its associated analytic functions h and g have the structures

$$h(z) = z + a_2z^2 + \cdots, \qquad g(z) = b_2z^2 + b_3z^3 + \cdots.$$

The mapping from $f \in C_H$ to $f_0 = \varphi \circ f \in C_H^0$ is inverted by $f = f_0 + \overline{b_1 \overline{f_0}}$.

A primary example of a function of class C_H^0 is

$$L(z) = \text{Re}\,\{\ell(z)\} + i\,\text{Im}\,\{k(z)\} = \frac{1}{2}[\ell(z) + k(z)] + \overline{\frac{1}{2}[\ell(z) - k(z)]},$$

obtained in Section 3.4 by vertical shearing of the half-plane mapping ℓ. Recall that L maps the disk harmonically onto the entire half-plane Re $\{w\} > -\frac{1}{2}$. Its coefficients are easily seen to be

$$a_n = \frac{n+1}{2} \qquad \text{and} \qquad b_n = -\frac{n-1}{2}, \qquad n = 1, 2, 3, \ldots.$$

In particular, L belongs to the class C_H^0. The following theorems are due to Clunie and Sheil-Small [1].

Theorem 1. *Each function $f \in C_H^0$ contains the full disk $|w| < \frac{1}{2}$ in its range $f(\mathbb{D})$.*

Theorem 2. *The coefficients of each function $f \in C_H^0$ satisfy the sharp inequalities*

$$|a_n| \leq \frac{n+1}{2}, \qquad |b_n| \leq \frac{n-1}{2}, \qquad and \qquad \big||a_n| - |b_n|\big| \leq 1$$

for $n = 2, 3, \ldots$. Equality occurs for the function L.

The proof of Theorem 1 depends on the *Herglotz representation* (*cf.* Duren [2], p. 22)

$$\varphi(z) = \int_0^{2\pi} \frac{e^{it} + z}{e^{it} - z} \, d\mu(t) + i\gamma$$

of a function φ analytic in $\mathbb{D}$ with positive real part. Here $d\mu$ is a positive measure and γ is a real constant. As a direct corollary, we have the following estimate of coefficients.

Lemma 1. *If $\varphi(z) = c_0 + c_1 z + \cdots$ is analytic with $\mathrm{Re}\,\{\varphi(z)\} > 0$ in $\mathbb{D}$, then $|c_n| \leq 2\,\mathrm{Re}\,\{c_0\}, n = 1, 2, \ldots$.*

Proof. The Herglotz representation shows that

$$c_n = 2 \int_0^{2\pi} e^{-int} \, d\mu(t), \qquad n = 1, 2, \ldots,$$

so that $|c_n| \leq 2\|\mu\| = 2\,\mathrm{Re}\,\{c_0\}$. ∎

Proof of Theorem 1. By hypothesis, the range $f(\mathbb{D})$ is convex. Thus, if $w \notin f(\mathbb{D})$, a suitable rotation will give

$$\mathrm{Re}\,\{e^{i\theta}[f(z) - w]\} > 0$$

for all $z \in \mathbb{D}$. But if $f = h + \overline{g}$, this says that $\mathrm{Re}\,\{\varphi(z)\} > 0$ for

$$\varphi(z) = e^{i\theta}[h(z) - w] + e^{-i\theta} g(z) = c_0 + c_1 z + \cdots,$$

where $c_0 = -e^{i\theta} w$ and $c_1 = e^{i\theta}$. Thus, Lemma 1 provides the inequality

$$1 = |e^{i\theta}| = |c_1| \leq 2|c_0| = 2|-e^{i\theta} w| = 2|w|,$$

or $|w| \geq \frac{1}{2}$. This proves the theorem. ∎

The proof of Theorem 2 is more difficult. It makes use of an intricate lemma, which is of some independent interest.

Lemma 2. *If* $f = h + \bar{g} \in C_H$, *then there exist angles* α *and* β *such that*

$$\text{Re}\,\{(e^{i\alpha}h'(z) + e^{-i\alpha}g'(z))(e^{i\beta} - e^{-i\beta}z^2)\} > 0$$

for all $z \in \mathbb{D}$.

Proof. By the approximation theorem of Section 2.6, it suffices to assume that f has a smooth extension to the boundary, so that $\lambda(t) = \frac{\partial}{\partial t}\{f(e^{it})\}$ is continuous and has increasing argument, with $\lambda(t + 2\pi) = \lambda(t)$. Thus, for each t there is a unique t^* with $t < t^* < t + 2\pi$ for which

$$\frac{\lambda(t^*)}{|\lambda(t^*)|} = -\frac{\lambda(t)}{|\lambda(t)|}.$$

Since t^* is a continuous function of t and $t^{**} = t + 2\pi$, it can be seen that $t_0^* = t_0 + \pi$ for some t_0. Then $\lambda(t)/\lambda(t_0)$ lies in the upper half-plane for $t_0 < t < t_0 + \pi$, whereas it lies in the lower half-plane for $t_0 + \pi < t < t_0 + 2\pi$. Since $\sin\theta = \frac{1}{2i}(e^{i\theta} - e^{-i\theta})$ is positive for $0 < \theta < \pi$ and negative for $\pi < \theta < 2\pi$, an equivalent statement is that

$$\text{Re}\,\left\{(e^{i(t_0-t)} - e^{i(t-t_0)})\frac{\lambda(t)}{\lambda(t_0)}\right\} \geq 0$$

for $t_0 \leq t \leq t_0 + 2\pi$. But a simple calculation shows that

$$\lambda(t) = ie^{it}h'(e^{it}) - ie^{-it}\overline{g'(e^{it})},$$

and the inequality takes the form

$$\text{Re}\,\left\{(e^{it_0} - e^{-it_0}e^{2it})\left(\frac{ih'(e^{it})}{\lambda(t_0)} - \frac{ig'(e^{it})}{\overline{\lambda(t_0)}}\right)\right\} \geq 0,$$

or

$$\text{Re}\,\{(e^{it_0} - e^{-it_0}z^2)(e^{i\alpha}h'(z) + e^{-i\alpha}g'(z))\} \geq 0$$

on the unit circle $|z| = 1$, where

$$e^{i\alpha} = \frac{i|\lambda(t_0)|}{\lambda(t_0)}.$$

The desired result now follows from the maximum principle for harmonic functions, completing the proof of Lemma 2. ∎

One additional fact will be required in the proof of Theorem 2. Recall that when f and g are analytic in $\mathbb{D}$, the function g is said to be *subordinate* to f, written $g \prec f$, if $g(z) = f(\omega(z))$ for some analytic function ω with $|\omega(z)| \leq |z|$ in $\mathbb{D}$. If f is *univalent* and $g(\mathbb{D}) \subset f(\mathbb{D})$ with $g(0) = f(0)$, an appeal to the Schwarz lemma shows that $g \prec f$. The following result is due to Rogosinski (cf. Duren [2], p. 195).

Lemma 3. *If* $g(z) = \sum_{n=1}^{\infty} b_n z^n$ *is analytic in* $\mathbb{D}$ *and* $g \prec f$ *for some* $f \in C$, *then* $|b_n| \leq 1$ *for* $n = 1, 2, \dots$.

Proof. Let Ω be the image of $\mathbb{D}$ under the mapping f, and let $\varepsilon_k = e^{2\pi i k/n}$ denote the nth roots of unity ($k = 1, 2, \dots, n$). Then since Ω is convex, it follows that

$$\varphi(z^n) = \frac{1}{n} \sum_{k=1}^{n} g(\varepsilon_k z) = b_n z^n + \cdots$$

lies in Ω for every $z \in \mathbb{D}$. This last expression is an analytic function of z^n, since $\sum_{k=1}^{n} (\varepsilon_k)^m = 0$ unless m is a multiple of n. Thus, $\varphi \prec f$, and $|b_n| = |\varphi'(0)| \leq |f'(0)| = 1$, by the Schwarz lemma. ∎

Proof of Theorem 2. It follows from Lemma 2 and the Herglotz representation that the Taylor coefficients of the function

$$e^{i\alpha} h'(z) + e^{-i\alpha} g'(z)$$

are dominated in modulus by the corresponding coefficients of the function

$$\frac{1+z}{1-z} \frac{1}{1-z^2} = \frac{1}{(1-z)^2}.$$

Integration now shows that the coefficients of $e^{i\alpha} h(z) + e^{-i\alpha} g(z)$ are dominated by those of $z/(1-z) = z + z^2 + z^3 + \cdots$. Thus,

$$\|a_n| - |b_n\| \leq |e^{i\alpha} a_n + e^{-i\alpha} b_n| \leq 1,$$

as the theorem asserts.

To obtain the other estimates, note that the dilatation $\omega(z) = g'(z)/h'(z)$ satisfies $|\omega(z)| < 1$ and $\omega(0) = 0$, so that $|\omega(z)| \leq |z|$ by the Schwarz lemma. Let

$$F(z) = (e^{i\alpha} h'(z) + e^{-i\alpha} g'(z))(e^{i\beta} - e^{-i\beta} z^2)$$

denote the function of Lemma 2, so that $\text{Re}\,\{F(z)\} > 0$ and $|F(0)| = 1$. Writing

$$g'(z) = \frac{\omega(z)}{e^{i\alpha} + e^{-i\alpha}\omega(z)} \frac{1}{e^{i\beta} - e^{-i\beta}z^2} F(z)$$

and appealing to Lemma 3, we see that the coefficients of $g'(z)$ are dominated by those of

$$\frac{z}{1-z} \frac{1}{1-z^2} \frac{1+z}{1-z} = \frac{z}{(1-z)^3} = \sum_{n=1}^{\infty} \frac{n(n+1)}{2} z^n.$$

In other words,

$$(n+1)|b_{n+1}| \le \frac{n(n+1)}{2}, \qquad n = 1, 2, \ldots,$$

or $|b_n| \le (n-1)/2$. The final assertion of Theorem 2 now follows by combining the other two estimates:

$$|a_n| \le \|a_n| - |b_n\| + |b_n| \le 1 + \frac{n-1}{2} = \frac{n+1}{2}.$$

This completes the proof. ∎

Theorem 2 leads also to sharp coefficient bounds for functions of class C_H.

Corollary. *The coefficients of each function $f \in C_H$ satisfy the sharp inequalities $|a_n| < n$ and $|b_n| < n, n = 2, 3, \ldots$.*

Proof. Recall that each function $f = h + \overline{g} \in C_H$ has the form $f = f_0 + \overline{b_1}\,\overline{f_0}$ for some function $f_0 \in C_H^0$, where $b_1 = g'(0)$ and $|b_1| < 1$. Thus, by Theorem 2,

$$|a_n| \le \frac{n+1}{2} + |b_1|\frac{n-1}{2} < n$$

and

$$|b_n| \le \frac{n-1}{2} + |b_1|\frac{n+1}{2} < n, \qquad n = 2, 3, \ldots.$$

To see that the bounds are sharp (although not attained for any function in S_H), we need only consider functions of the form

$$f(z) = L(z) - b\,\overline{L(z)}, \qquad 0 < b < 1,$$

where L is the standard harmonic mapping of the disk onto the half-plane $\text{Re}\,\{w\} > -\frac{1}{2}$. ∎

For the full class of normalized harmonic mappings, not necessarily convex, the sharp analogues of Theorems 1 and 2 are not known. We shall return to these questions in Chapter 5, where some partial results will be developed.

Other results on convex harmonic mappings may be found in the literature, for instance in the papers by Abu-Muhanna and Schober [1], Dorff [2,3], and Goodloe [1].

3.7. Failure of the Radó–Kneser–Choquet Theorem in $\mathbb{R}^3$

The Radó–Kneser–Choquet theorem says that any prescribed homeomorphism of the unit circle onto a closed convex curve in the finite plane extends to a (univalent) harmonic mapping of the unit disk onto the domain inside that curve. It is natural to ask whether the result generalizes to higher dimensions. In the absence of the Riemann mapping theorem, it is not altogether clear how the question should be formulated, but it seems reasonable to select the unit ball as the domain of definition. Given a homeomorphism of the unit sphere onto the boundary of a bounded convex domain in $\mathbb{R}^n$, the question is then whether the harmonic extension to the ball is necessarily univalent.

The answer is NO. In 1994, Richard Laugesen [1] constructed a homeomorphism of the unit sphere in $\mathbb{R}^3$ onto itself, whose Poisson extension to a vector-valued harmonic function fails to be univalent in the ball. The idea of the construction is to use what Laugesen calls a "tennis-ball homeomorphism." This can be viewed as deforming the sphere by moving a large part of the northern hemisphere into the southern hemisphere, and vice versa, so that the equator is transformed to a curve that resembles the seam of a tennis ball. If this is done symmetrically, the resulting harmonic extension will map the polar axis of the sphere onto itself but will produce a folding near the center.

Here are the details. Let $x = (x, y, z)$ and $u = (u, v, w)$ denote points in $\mathbb{R}^3$, and let $|x|^2 = x^2 + y^2 + z^2$. For any continuous vector-valued function $x = g(u)$ on the unit sphere

$$S^2 = \{u \in \mathbb{R}^3 : |u| = 1\},$$

the harmonic extension to the unit ball

$$B^3 = \{x \in \mathbb{R}^3 : |x| < 1\}$$

is given by the Poisson formula

$$f(x) = \frac{1}{4\pi} \int_{S^2} \frac{1 - |x|^2}{|x - u|^3}\, g(u)\, d\sigma(u), \qquad x \in B^3,$$

where $d\sigma$ denotes the element of surface area. In terms of spherical coordinates

$$u = \sin\varphi \cos\theta, \quad v = \sin\varphi \sin\theta, \quad w = \cos\varphi,$$

where $0 \leq \theta \leq 2\pi$ and $0 \leq \varphi \leq \pi$, the formula can be written as

$$f(x) = \frac{1}{4\pi} \int_0^\pi \int_0^{2\pi} \frac{1 - |x|^2}{|x - u|^3} g(\theta, \varphi) \sin\varphi \, d\theta \, d\varphi.$$

Now let $g = (g_1, g_2, g_3)$ be a homeomorphism of the unit sphere onto itself, chosen to fix the north and south poles and to have the symmetries $g_1(\theta + \pi, \varphi) = -g_1(\theta, \varphi)$, $g_2(\theta + \pi, \varphi) = -g_2(\theta, \varphi)$, and $g_3(\theta + \frac{\pi}{2}, \pi - \varphi) = -g_3(\theta, \varphi)$. Then, with the notation

$$G(\varphi) = \frac{1}{2\pi} \int_0^{2\pi} g(\theta, \varphi) \, d\theta = (G_1(\varphi), G_2(\varphi), G_3(\varphi)),$$

we find $G_1(\varphi) = G_2(\varphi) = 0$ and $G_3(\pi - \varphi) = -G_3(\varphi)$ for $0 \leq \varphi \leq \pi$. For a fixed constant c ($0 < c < \frac{\pi}{4}$) to be specified later, it will be further required that $G_3(\varphi) < -\frac{1}{2}$ for all φ in the interval $c \leq \varphi \leq \frac{\pi}{2} - c$. Intuitively, this requirement ensures that the homeomorphism g transforms a substantial portion of the northern hemisphere well into the southern hemisphere, and vice versa (by symmetry).

Since g is constructed to fix the poles $k = (0,0,1)$ and $-k$, its harmonic extension f will do the same: $f(k) = k$ and $f(-k) = -k$. To investigate the behavior of $f = (f_1, f_2, f_3)$ on the polar axis, let $-1 < z < 1$ and write

$$f(zk) = \frac{1}{4\pi} \int_0^\pi \int_0^{2\pi} \frac{1 - z^2}{[\sin^2\varphi + (z - \cos\varphi)^2]^{3/2}} g(\theta, \varphi) \sin\varphi \, d\theta \, d\varphi$$

$$= \frac{1}{2} \int_0^\pi \frac{1 - z^2}{[1 - 2z\cos\varphi + z^2]^{3/2}} G(\varphi) \sin\varphi \, d\varphi.$$

Thus, $f_1(zk) = f_2(zk) = 0$, so that f maps the polar axis onto itself. The height of the image is found to be

$$f_3(zk) = \frac{1 - z^2}{2} \int_0^{\pi/2} H(z, \varphi) \, G_3(\varphi) \sin\varphi \, d\varphi,$$

where

$$H(z, \varphi) = \frac{1}{[1 - 2z\cos\varphi + z^2]^{3/2}} - \frac{1}{[1 + 2z\cos\varphi + z^2]^{3/2}},$$

in view of the relation $G_3(\pi - \varphi) = -G_3(\varphi)$. In particular, $f_3(0) = 0$; the harmonic extension fixes the center of the ball. Observe now that $H(z, \varphi)$ is continuous in $0 \leq \varphi \leq \frac{\pi}{2}$ for each fixed $z > 0$, and $H(z, \varphi) > 0$ for $0 \leq \varphi < \frac{\pi}{2}$.

Given $z > 0$, we can therefore choose a sufficiently small positive constant $c < \frac{\pi}{4}$ such that $I \geq 4J > 0$, where

$$I = \frac{1 - z^2}{2} \int_c^{\pi/2 - c} H(z, \varphi) \sin \varphi \, d\varphi$$

and

$$J = \frac{1 - z^2}{2} \left\{ \int_0^c H(z, \varphi) \sin \varphi \, d\varphi + \int_{\pi/2 - c}^{\pi/2} H(z, \varphi) \sin \varphi \, d\varphi \right\}.$$

But by construction, $G_3(\varphi) < -\frac{1}{2}$ for $c \leq \varphi \leq \frac{\pi}{2} - c$ and $G_3(\varphi) \leq 1$ for $0 \leq \varphi \leq \frac{\pi}{2}$, so we conclude that

$$f_3(z\mathbf{k}) \leq -\frac{1}{2} I + J \leq -J < 0.$$

Since f maps the polar axis onto itself, with $f(\mathbf{k}) = \mathbf{k}$ and $f(\mathbf{0}) = \mathbf{0}$, the fact that $f(z\mathbf{k})$ is in the lower hemisphere implies that the restriction of f to the upper half of the polar axis is not univalent; it exhibits a folding. In particular, the harmonic extension of the homeomorphism g is not univalent in the ball.

This example demonstrates the failure of the Radó–Kneser–Choquet theorem in $\mathbb{R}^3$, even for homeomorphisms of the sphere S^2 onto itself. Laugesen [1] carries out the construction in more explicit detail and generalizes it to homeomorphisms of the sphere S^{n-1} onto itself, thus showing that the theorem fails in $\mathbb{R}^n$ for any $n > 2$. A striking feature of his construction is that the homeomorphism can be chosen arbitrarily close to the identity in the uniform norm, yet its harmonic extension to the ball B^n may fail to be univalent.

In a related construction for $\mathbb{R}^3$, Melas [1] showed that the harmonic extension of a homeomorphism of S^2 need not be a diffeomorphism. Liu and Liao [1] followed the approach of Melas and claimed to construct a homeomorphism of S^2 whose harmonic extension to the ball is not univalent, but their argument appears to be flawed because the mapping of S^2 they display need not be injective.

4

Harmonic Self-Mappings of the Disk

The main focus of this chapter will be the harmonic mappings of the unit disk onto itself and the closely related mappings of the disk onto inscribed polygons. Other topics include the sharp form of Heinz' inequality, coefficient estimates, and a version of the Schwarz lemma.

4.1. Representation by Radó–Kneser–Choquet Theorem

One consequence of the Radó–Kneser–Choquet theorem is the existence of many harmonic mappings of the disk onto a given convex region. In contrast to the rigid behavior of conformal mappings, which are determined completely by the images of three boundary points, harmonic mappings allow the boundary correspondence to be prescribed at every point with only mild restrictions. The greater flexibility is sometimes an advantage, but it has the obvious disadvantage that a harmonic mapping is in no way determined by its range.

Even the harmonic mappings of the disk onto itself form a very large family. According to the Radó–Kneser–Choquet theorem and the results on boundary behavior of harmonic mappings (Section 3.3), the harmonic self-mappings of the disk correspond to the continuous monotonic self-mappings of the circle. More precisely, if $\theta = \theta(t)$ is a continuous nondecreasing function on the interval $[0, 2\pi]$, with $\theta(2\pi) - \theta(0) = 2\pi$, then

$$f(z) = \frac{1}{2\pi} \int_0^{2\pi} \frac{1 - |z|^2}{|e^{it} - z|^2} e^{i\theta(t)} dt$$

provides a sense-preserving harmonic mapping of $\mathbb{D}$ onto $\mathbb{D}$; conversely, each such mapping is continuous in $\overline{\mathbb{D}}$ and has a boundary function $f(e^{it}) = e^{i\theta(t)}$ where $\theta(t)$ is a function with the given properties. More generally, any nondecreasing function $\theta(t)$ on $[0, 2\pi)$ with $\theta(2\pi-) - \theta(0) \leq 2\pi$ gives rise in this way to a harmonic mapping of $\mathbb{D}$ *into* $\mathbb{D}$, provided that the points $e^{i\theta(t)}$ do not lie on a line for almost every t in $[0, 2\pi)$.

For example, if $\theta(t)$ is taken to have the values

$$\theta(t) = \begin{cases} 0, & 0 \le t \le 2\pi/3 \\ 2\pi/3, & 2\pi/3 < t \le 4\pi/3 \\ 4\pi/3, & 4\pi/3 < t < 2\pi, \end{cases}$$

then f is a harmonic mapping of $\mathbb{D}$ onto the interior of the equilateral triangle inscribed in $\mathbb{T}$ with vertices at the cube roots of unity: 1, $e^{2\pi i/3}$, and $e^{4\pi i/3}$. In fact, f is a linear combination of harmonic measures of the three intervals. The given function $\theta(t)$ has discontinuities at $t = 0, 2\pi/3$, and $4\pi/3$. It can be approximated by a sequence of continuous piecewise-linear functions $\theta_n(t)$ defined by replacing the graph of $\theta(t)$ with a linear function in each of the intervals of length $2/n$ centered (modulo 2π) at 0, $2\pi/3$, and $4\pi/3$. For $n = 2, 3, \ldots$, the corresponding harmonic function f_n will map $\mathbb{D}$ univalently onto itself. It follows from the Poisson representation that $f_n(z)$ converges to $f(z)$ uniformly on each compact subset of $\mathbb{D}$ as n tends to infinity. Thus, we have constructed a sequence of harmonic mappings of $\mathbb{D}$ onto $\mathbb{D}$ that converges locally uniformly to a harmonic mapping of $\mathbb{D}$ onto an inscribed triangle. This example demonstrates the failure of the Carathéodory convergence theorem (see Section 1.5) for harmonic mappings, even when the limit function is univalent.

On the other hand, the limit function need not be univalent. For instance, let

$$\theta(t) = \begin{cases} 0, & 0 \le t \le \pi \\ \pi, & \pi < t < 2\pi. \end{cases}$$

Then the corresponding harmonic function f maps $\mathbb{D}$ onto the real segment $-1 < u < 1$. A straightforward calculation leads to the formula

$$f(z) = \frac{2}{\pi} \arg\left\{\frac{1+z}{1-z}\right\}.$$

Geometrically, the formula shows that f projects each circular arc through the points 1 and -1 onto a single point of the real segment $(-1,1)$. It also shows that $f(0) = 0$. The derivatives are found to be

$$f_z(z) = \frac{2}{\pi i} \frac{1}{1 - z^2} = \overline{f_{\bar{z}}(z)},$$

which implies that the Jacobian is

$$J_f(z) = |f_z(z)|^2 - |f_{\bar{z}}(z)|^2 \equiv 0,$$

while the dilatation is

$$\omega(z) = \overline{f_{\bar{z}}(z)}/f_z(z) \equiv 1.$$

If $\theta(t)$ is approximated as before by continuous piecewise-linear functions $\theta_n(t)$, the Poisson formula gives a sequence $\{f_n\}$ of (univalent) harmonic mappings of $\mathbb{D}$ onto itself that converge locally uniformly to the nonunivalent harmonic function f. Observe in particular that $f_n(0) = 0$ and the Jacobian of f_n is positive at the origin (by Lewy's theorem), yet $J_{f_n}(0) \to J_f(0) = 0$.

Thus, although for all sense-preserving harmonic mappings f of $\mathbb{D}$ onto $\mathbb{D}$ with $f(0) = 0$ the Jacobians are positive at the origin, they do not have a positive lower bound there. In contrast, the Heinz lemma (see Section 2.3) asserts that for all such mappings the expressions $|f_z(0)|^2 + |f_{\bar{z}}(0)|^2$ are bounded below by a positive constant.

4.2. Mappings onto Regular Polygons

Generalizing the examples given in the preceding section, we shall now construct a canonical harmonic mapping of the disk onto the domain inside a regular n-gon with vertices at the nth roots of unity. Fix an integer $n = 3, 4, \ldots$, and let $\alpha = e^{2\pi i/n}$ be the primitive nth root of unity. Define a step function $\theta = \theta(t)$ by the formula

$$\theta(t) = 2k\pi/n, \qquad (2k-1)\pi/n < t < (2k+1)\pi/n,$$

for $k = 0, 1, \ldots, n-1$. Then $e^{i\theta(t)} = \alpha^k$ when e^{it} lies on the circular arc centered at α^k, of length $2\pi/n$. The corresponding harmonic mapping f, the Poisson integral of the given boundary function $e^{i\theta(t)}$, is a linear combination of harmonic measures of subarcs centered at the nth roots of unity, with coefficients running through the same nth roots of unity.

For convenience, let $\beta = \sqrt{\alpha} = e^{i\pi/n}$. Then the kth subarc extends from β^{2k-1} to β^{2k+1}, and its harmonic measure (see Section 1.4) is

$$\omega_k(z) = \frac{1}{\pi} \arg \left\{ \frac{z - \beta^{2k+1}}{z - \beta^{2k-1}} \right\} - \frac{1}{n}.$$

Thus, the harmonic mapping takes the form

$$f(z) = \sum_{k=0}^{n-1} \alpha^k \omega_k(z) = \frac{1}{\pi} \sum_{k=0}^{n-1} \alpha^k \arg \left\{ \frac{z - \beta^{2k+1}}{z - \beta^{2k-1}} \right\},$$

since

$$1 + \alpha + \alpha^2 + \cdots + \alpha^{n-1} = 0.$$

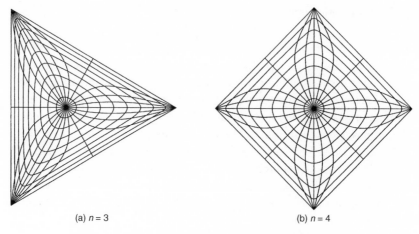

Figure 4.1. Image of canonical mapping onto regular n-gon.

It follows from the general form of Choquet's proof, as given in Section 3.2, that f is univalent in $\mathbb{D}$ and it provides a harmonic mapping onto the region inside the regular n-gon with vertices $1, \alpha, \alpha^2, \ldots, \alpha^{n-1}$. The image of the disk under this mapping f is displayed in Figure 4.1 for $n = 3$ and 4. Note that the images of radial segments meet the boundary at one of the nth roots of unity, as expected. The only exceptions are radial segments terminating at a jump point of the boundary function.

Our next aim is to find the dilatation of this mapping f. In order to calculate the derivatives $\partial f/\partial z$ and $\partial f/\partial \overline{z}$, it is convenient to write

$$f(z) = \frac{1}{2\pi i} \sum_{k=0}^{n-1} \alpha^k \left\{ \log \frac{z - \beta^{2k+1}}{z - \beta^{2k-1}} - \overline{\log \frac{z - \beta^{2k+1}}{z - \beta^{2k-1}}} \right\}.$$

Differentiation gives

$$f_z(z) = \frac{1}{2\pi i} \sum_{k=0}^{n-1} \alpha^k \left\{ \frac{1}{z - \beta^{2k+1}} - \frac{1}{z - \beta^{2k-1}} \right\}$$

$$= \frac{1 - \alpha}{2\pi i} \sum_{k=0}^{n-1} \frac{\alpha^k}{z - \beta^{2k+1}},$$

while

$$\overline{f_{\overline{z}}(z)} = \frac{1 - \overline{\alpha}}{2\pi i} \sum_{k=0}^{n-1} \frac{\overline{\alpha}^k}{z - \beta^{2k+1}}.$$

The two identities

$$\sum_{k=0}^{n-1} \frac{\alpha^k}{z - \beta^{2k+1}} = \frac{n\beta^{n-1}}{z^n + 1}$$

and

$$\sum_{k=0}^{n-1} \frac{\overline{\alpha}^k}{z - \beta^{2k+1}} = \frac{n\beta z^{n-1}}{z^n + 1}$$

may now be used. To verify these partial-fraction expansions, simply multiply both sides of the equations by $(z - \beta^{2k+1})$ and let z tend to β^{2k+1}. The dilatation of f is therefore

$$\omega(z) = \overline{f_{\overline{z}}(z)}/f_z(z) = \frac{1 - \overline{\alpha}}{1 - \alpha} \frac{n\beta z^{n-2}}{n\beta^{n-1}} = z^{n-2},$$

since $\beta^n = -1$.

Finally, we shall compute the Fourier coefficients of the mapping f just constructed. First note that

$$f(0) = \frac{1}{\pi} \sum_{k=0}^{n-1} \alpha^k \arg \{\alpha\} = 0.$$

The canonical decomposition is $f = h + \overline{g}$, where

$$h(z) = \frac{1}{2\pi i} \sum_{k=0}^{n-1} \alpha^k \log \frac{z - \beta^{2k+1}}{z - \beta^{2k-1}} = \sum_{m=1}^{\infty} a_m z^m$$

and

$$g(z) = \frac{1}{2\pi i} \sum_{k=0}^{n-1} \overline{\alpha}^k \log \frac{z - \beta^{2k+1}}{z - \beta^{2k-1}} = \sum_{m=1}^{\infty} b_m z^m.$$

To calculate the coefficients a_m and b_m, consider first the Taylor expansion

$$\log \frac{z - \beta^{2k+1}}{z - \beta^{2k-1}} = \log \alpha + \log (1 - \overline{\beta}^{2k+1} z) - \log (1 - \overline{\beta}^{2k-1} z)$$

$$= \log \alpha - \sum_{m=1}^{\infty} \frac{1}{m} (\overline{\beta}^{2k+1} z)^m + \sum_{m=1}^{\infty} \frac{1}{m} (\overline{\beta}^{2k-1} z)^m.$$

This formula shows that

$$a_m = \frac{\beta^m - \overline{\beta}^m}{2m\pi i} \sum_{k=0}^{n-1} \alpha^k \overline{\alpha}^{km} = \frac{\sin m\pi/n}{\pi m} \sum_{k=0}^{n-1} \alpha^k \overline{\alpha}^{km}.$$

But

$$\sum_{k=0}^{n-1} \alpha^k \, \overline{\alpha}^{km} = \sum_{k=0}^{n-1} \alpha^{(1-m)k} = \begin{cases} 0, & \alpha^{m-1} \neq 1 \\ n, & \alpha^{m-1} = 1. \end{cases}$$

Thus,

$$a_m = \frac{n}{\pi m} \sin \frac{m\pi}{n}, \qquad m = 1, n+1, 2n+1, \ldots,$$

and $a_m = 0$ otherwise.

A similar calculation shows that

$$b_m = \frac{\sin m\pi/n}{\pi m} \sum_{k=0}^{n-1} \overline{\alpha}^{(m+1)k},$$

whence

$$b_m = \frac{n}{\pi m} \sin \frac{m\pi}{n}, \qquad m = n-1, 2n-1, \ldots,$$

and $b_m = 0$ otherwise. This is consistent with the previous information that $\omega(z) = z^{n-2}$, or $g'(z) = z^{n-2}h'(z)$, as a comparison of coefficients shows.

For more general mappings onto regions bounded by convex polygons, not necessarily regular, it turns out that the dilatation is always a finite Blaschke product of degree $n - 2$, where n is the number of vertices. This striking result, due to Sheil-Small, will be developed in the next section.

4.3. Arbitrary Convex Polygons

The last section dealt with a standard harmonic mapping $f = h + \overline{g}$ of the unit disk onto a regular n-gon inscribed in the unit circle. Its dilatation $\omega = g'/h'$ was found to have the form $\omega(z) = z^{n-2}$. We now consider a remarkable generalization to harmonic mappings onto arbitrary convex polygons with n vertices, generated as before by piecewise constant boundary functions. Following Sheil-Small [2], we shall calculate the dilatation and show it to be a finite Blaschke product with precisely $n - 2$ factors. In Section 7.4 we address the question whether *every* finite Blaschke product can arise as a dilatation of some harmonic mapping of this type.

Recall first that a *finite Blaschke product* of *degree* n, or with n factors, is a function of the form

$$B(z) = \gamma \prod_{k=1}^{n} \frac{z - z_k}{1 - \overline{z}_k z},$$

where $|\gamma| = 1$ and $|z_k| < 1$ for $k = 1, 2, \ldots, n$. Such a function B is analytic in the closed unit disk, with $|B(z)| < 1$ for $|z| < 1$ and $|B(z)| = 1$ for $|z| = 1$. The points z_k need not be distinct.

Now let Ω be a convex polygon with n distinct vertices $\alpha_1, \alpha_2, \ldots, \alpha_n$ taken in counterclockwise order around the boundary. Choose any partition $0 = t_0 < t_1 < \cdots < t_n = 2\pi$ of the interval $[0, 2\pi]$ and define the step function $\varphi(e^{it}) = \alpha_k$ for $t_{k-1} < t < t_k$, $k = 1, 2, \ldots, n$. According to the strong form of the Radó–Kneser–Choquet theorem (see remark at end of Section 3.2), the harmonic extension

$$f(z) = \frac{1}{2\pi} \int_0^{2\pi} \frac{1 - |z|^2}{|e^{it} - z|^2} \varphi(e^{it}) \, dt$$

is univalent in the unit disk and maps it onto Ω. The mapping has the canonical representation $f = h + \overline{g}$, where h and g are analytic and $g(0) = 0$. By writing the Poisson kernel as

$$\frac{1 - |z|^2}{|e^{it} - z|^2} = \frac{e^{it}}{e^{it} - z} + \frac{\overline{z}}{e^{-it} - \overline{z}},$$

we see that

$$h(z) = \frac{1}{2\pi} \int_0^{2\pi} \frac{e^{it}}{e^{it} - z} \varphi(e^{it}) \, dt$$

and

$$g(z) = \frac{1}{2\pi} \int_0^{2\pi} \frac{z}{e^{it} - z} \overline{\varphi(e^{it})} \, dt.$$

Taking derivatives and inserting the definition of the step function $\varphi(e^{it})$, we arrive at the formulas

$$h'(z) = \frac{1}{2\pi i} \sum_{k=1}^n \alpha_k \left(\frac{1}{z - \zeta_k} - \frac{1}{z - \zeta_{k-1}} \right)$$

and

$$g'(z) = \frac{1}{2\pi i} \sum_{k=1}^n \overline{\alpha_k} \left(\frac{1}{z - \zeta_k} - \frac{1}{z - \zeta_{k-1}} \right),$$

where $\zeta_k = e^{it_k}$. In more compact notation,

$$h'(z) = \sum_{k=1}^n \frac{c_k}{z - \zeta_k} \quad \text{and} \quad g'(z) = -\sum_{k=1}^n \frac{\overline{c_k}}{z - \zeta_k},$$

where $c_k = \frac{1}{2\pi i}(\alpha_k - \alpha_{k+1})$ and $\alpha_{n+1} = \alpha_1$. Since $|\zeta_k| = 1$ and $\sum_{k=1}^{n} c_k = 0$, these last formulas lead to the structural relation

$$\overline{h'(1/\bar{z})} = z^2 g'(z), \qquad \text{or} \qquad \overline{g'(1/\bar{z})} = z^2 h'(z).$$

Indeed,

$$\overline{h'(1/\bar{z})} - z^2 g'(z) = -z \sum_{k=1}^{n} \frac{\zeta_k \overline{c_k}}{z - \zeta_k} + z^2 \sum_{k=1}^{n} \frac{\overline{c_k}}{z - \zeta_k} = z \sum_{k=1}^{n} \overline{c_k} = 0.$$

The rational functions h' and g' can be expressed as

$$h'(z) = \frac{P(z)}{S(z)} \qquad \text{and} \qquad g'(z) = \frac{Q(z)}{S(z)},$$

where

$$S(z) = \prod_{k=1}^{n} (z - \zeta_k)$$

and P and Q are polynomials of degree at most $n - 2$, since $\sum_{k=1}^{n} c_k = 0$. But

$$\overline{S(1/\bar{z})} = (-1)^n \left(\prod_{k=1}^{n} \overline{\zeta_k} \right) z^{-n} S(z),$$

so the structural relation between h' and g' gives

$$z^{n-2} \, \overline{Q(1/\bar{z})} = (-1)^n \left(\prod_{k=1}^{n} \overline{\zeta_k} \right) P(z),$$

and the same formula holds with P and Q interchanged. But $h'(z) \neq 0$ in $\mathbb{D}$, since f is orientation-preserving, and so $P(z) \neq 0$ in $\mathbb{D}$. In particular, $P(0) \neq 0$, and so the last equation shows that Q has degree $n - 2$. Thus, Q has a factorization of the form

$$Q(z) = C z^m \prod_{k=1}^{n-m-2} (z - z_k), \qquad z_k \neq 0,$$

where $|z_k| \leq 1$ because $P(z) \neq 0$ in $\mathbb{D}$. Hence,

$$P(z) = \beta C \prod_{k=1}^{n-m-2} (1 - \overline{z_k} z), \qquad |\beta| = 1, \qquad |z_k| \leq 1.$$

Consequently, the mapping f has dilatation

$$\omega(z) = \frac{g'(z)}{h'(z)} = \frac{Q(z)}{P(z)} = \gamma z^m \prod_{k=1}^{n-m-2} \frac{z - z_k}{1 - \overline{z_k} z}, \qquad |\gamma| = 1,$$

a Blaschke product of degree at most $n - 2$.

To show that the degree of ω is *precisely* $n - 2$, it must be proved that $|z_k| < 1$ for $k = 1, 2, \ldots, n - m - 2$ or, equivalently, that P has no zeros on the unit circle. Observe that $P(\zeta_k) \neq 0$ for $k = 1, 2, \ldots, n$, since $h' = P/S$ and h' has a simple pole at each point ζ_k. Thus, it is enough to show that $|h'(z)|$ has a positive lower bound in $\mathbb{D}$. For this we need the following lemma.

Lemma. *For any finite Blaschke product $B(z)$, the ratio $(1 - |B(z)|)/(1 - |z|)$ is bounded in the unit disk.*

Deferring the proof of the lemma, let us use it to show that $|h'(z)|$ has a positive lower bound in $\mathbb{D}$ and, hence, that P and Q have no (common) zeros on $\mathbb{T}$. Since f is a convex harmonic mapping, it follows that $h + \alpha g$ is univalent in $\mathbb{D}$ for each unimodular constant α (see Section 3.4, Corollary to Theorem 2). Also,

$$|h'(0) + \alpha g'(0)| \geq |h'(0)| - |g'(0)| = \delta > 0,$$

since f is sense-preserving. Thus by Koebe's distortion theorem,

$$|h'(z) + \alpha g'(z)| \geq \delta \frac{1 - |z|}{(1 + |z|)^3} \geq \frac{\delta}{8}(1 - |z|)$$

for each $z \in \mathbb{D}$. With appropriate choice of $\alpha = \alpha(z)$, it follows that

$$(1 - |\omega(z)|)|h'(z)| = |h'(z)| - |g'(z)| \geq \frac{\delta}{8}(1 - |z|),$$

and the lemma shows that $|h'(z)|$ has a positive lower bound in $\mathbb{D}$.

Proof of Lemma. The function $B(z)$ is analytic in $\overline{\mathbb{D}}$ and thus satisfies a Lipschitz condition there. Hence, for $0 < |z| < 1$ and $\zeta = z/|z|$,

$$\frac{1 - |B(z)|}{1 - |z|} \leq \left| \frac{B(\zeta) - B(z)}{\zeta - z} \right| \leq M,$$

since $|B(\zeta)| = 1$. ∎

Finally, it is of interest to calculate the power-series coefficients of h and g. With the usual notation

$$h(z) = \sum_{m=0}^{\infty} a_m z^m \qquad \text{and} \qquad g(z) = \sum_{m=1}^{\infty} b_m z^m,$$

the integral formulas for h and g give

$$a_m = \frac{1}{2\pi} \int_0^{2\pi} e^{-imt} \, \varphi(e^{it}) \, dt, \qquad b_m = \frac{1}{2\pi} \int_0^{2\pi} e^{-imt} \, \overline{\varphi(e^{it})} \, dt.$$

Thus,

$$a_0 = \frac{1}{2\pi} \sum_{k=1}^{n} \alpha_k (t_k - t_{k-1})$$

and

$$a_m = \frac{1}{2\pi} \sum_{k=1}^{n} \alpha_k \int_{t_{k-1}}^{t_k} e^{-imt} \, dt = \frac{1}{2\pi i m} \sum_{k=1}^{n} \alpha_k \left(\overline{\zeta_{k-1}}^{\,m} - \overline{\zeta_k}^{\,m} \right)$$

for $m = 1, 2, \ldots$. Similarly,

$$b_m = \frac{1}{2\pi i m} \sum_{k=1}^{n} \overline{\alpha_k} \left(\overline{\zeta_{k-1}}^{\,m} - \overline{\zeta_k}^{\,m} \right), \qquad m = 1, 2, \ldots .$$

4.4. Sharp Form of Heinz's Inequality

Heinz's inequality asserts that $|f_z(0)|^2 + |f_{\bar{z}}(0)|^2 \geq c$ for all harmonic mappings f of the unit disk onto itself with $f(0) = 0$, where $c > 0$ is an absolute constant. A version of Heinz's original proof, leading to the value $c = 0.1788 \ldots$, was presented in Section 2.3. We now turn to the sharp form of the inequality with the constant $c = 27/4\pi^2 = 0.6839 \ldots$, established by Richard Hall [2] in 1982.

Let us first observe that no better estimate is possible. The value $c = 27/4\pi^2$ is attained by the canonical mapping

$$f(z) = \frac{1}{\pi} \sum_{k=0}^{2} \beta^{2k} \arg \left\{ \frac{z - \beta^{2k+1}}{z - \beta^{2k-1}} \right\}, \qquad \beta = e^{i\pi/3},$$

of the disk onto an inscribed equilateral triangle, as developed in Section 4.2. According to formulas already derived,

$$f_z(0) = a_1 = \frac{3}{\pi} \sin \frac{\pi}{3} = \frac{3\sqrt{3}}{2\pi} \qquad \text{and} \qquad f_{\bar{z}}(0) = \overline{b_1} = 0.$$

Thus, for the triangle mapping,

$$|f_z(0)|^2 + |f_{\bar{z}}(0)|^2 = \frac{27}{4\pi^2}.$$

This function f does not map the unit disk onto itself, but as a consequence of the Radó–Kneser–Choquet theorem (Section 3.1) it can be approximated uniformly on compact sets by harmonic mappings of the disk onto itself that preserve the origin. Thus, Heinz' inequality is false for each $c > 27/4\pi^2$.

The sense-preserving harmonic mappings of the disk onto itself can be represented as Poisson extensions of boundary functions $f(e^{it}) = e^{i\theta(t)}$, where $\theta(t)$ is a continuous nondecreasing function with $\theta(2\pi) - \theta(0) = 2\pi$ (see Sections 3.1 and 3.3). It will be convenient to extend $\theta(t)$ to the whole real line by the relation $\theta(t + 2\pi) = \theta(t) + 2\pi$. The coefficients $a_1 = f_z(0)$, $b_1 = \overline{f_{\bar{z}}(0)}$, and $a_0 = f(0)$ have an alternate interpretation as Fourier coefficients of the periodic function $e^{i\theta(t)}$, and so Heinz's lemma can be viewed as a statement about Fourier series, without reference to harmonic mappings. Following Hall [2], we shall prove the following theorem.

Theorem. *The coefficients of every harmonic mapping of the unit disk onto itself satisfy the inequality*

$$|a_1|^2 + \frac{3\sqrt{3}}{\pi}|a_0|^2 + |b_1|^2 \geq \frac{27}{4\pi^2}.$$

The lower bound $27/4\pi^2$ is the best possible.

Proof. There is no generality lost in restricting attention to sense-preserving mappings. Then the canonical coefficients a_n and b_n may be regarded as Fourier coefficients of the boundary function $e^{i\theta(t)}$, where $\theta(t)$ is a continuous nondecreasing function with $\theta(t + 2\pi) = \theta(t) + 2\pi$. An application of Parseval's relation leads to the expression

$$\frac{1}{2\pi}\int_0^{2\pi} e^{i[\theta(s+t)-\theta(s-t)]}\, ds = |a_0|^2 + \sum_{n=1}^{\infty}\left(|a_n|^2 + |b_n|^2\right)e^{2int}$$

for arbitrary $t \in \mathbb{R}$. Taking real parts, we arrive at the formula

$$1 - 2J(t) = |a_0|^2 + \sum_{n=1}^{\infty}\left(|a_n|^2 + |b_n|^2\right)\cos 2nt,$$

where

$$J(t) = \frac{1}{2\pi}\int_0^{2\pi} \sin^2\left(\frac{\theta(s+t) - \theta(s-t)}{2}\right)ds.$$

Now let $N(t)$ be the even function with period $\frac{\pi}{3}$ that has the values $\cos^2\left(\frac{\pi}{3} + t\right)$ in the interval $0 \le t \le \frac{\pi}{6}$. Let

$$M(t) = \cos^2 t - N(t), \qquad 0 \le t \le \frac{\pi}{2}.$$

Then

$$M(t) = \begin{cases} \cos^2 t - \cos^2\left(\dfrac{\pi}{3} + t\right), & 0 \le t \le \dfrac{\pi}{6}, \\[2mm] \cos^2 t - \cos^2\left(\dfrac{2\pi}{3} - t\right), & \dfrac{\pi}{6} \le t \le \dfrac{\pi}{3}, \\[2mm] 0, & \dfrac{\pi}{3} \le t \le \dfrac{\pi}{2}. \end{cases}$$

On the basis of the Fourier expansion

$$N(t) = \frac{1}{2} - \frac{3\sqrt{3}}{4\pi} + \frac{3\sqrt{3}}{2\pi} \sum_{n=1}^{\infty} \frac{1}{9n^2 - 1} \cos 6nt$$

and the formula for $1 - 2J(t)$ derived earlier, one calculates

$$\frac{8}{\pi} \int_0^{\pi/2} M(t)(1 - 2J(t))\, dt = |a_1|^2 + \frac{3\sqrt{3}}{\pi}|a_0|^2 + |b_1|^2$$

$$- \frac{3\sqrt{3}}{\pi} \sum_{n=1}^{\infty} \frac{1}{9n^2 - 1}\left(|a_{3n}|^2 + |b_{3n}|^2\right) \le |a_1|^2 + \frac{3\sqrt{3}}{\pi}|a_0|^2 + |b_1|^2.$$

Since

$$\frac{8}{\pi} \int_0^{\pi/2} M(t)\, dt = \frac{3\sqrt{3}}{\pi},$$

it must now be shown that

$$(\star) \qquad \frac{16}{\pi} \int_0^{\pi/2} M(t) J(t)\, dt \le \frac{3\sqrt{3}}{\pi} - \frac{27}{4\pi^2}.$$

This will be achieved with the help of the following lemmas.

Lemma 1. *Let $t_1, t_2, \ldots, t_n$ be nonnegative numbers with sum $t_1 + t_2 + \cdots + t_n = \pi$. Then $J(t_1) + J(t_2) + \cdots + J(t_n) \le \frac{9}{4}$.*

Lemma 2. *If a continuous function $P(t)$ is nonnegative and nonincreasing in the interval $0 \le t \le \frac{\pi}{2}$, then*

$$\int_0^{\pi/2} P(t) J(t)\, dt \le \frac{9}{4\pi} \int_0^{\pi/2} t P(t)\, dt.$$

Deferring the proofs of the lemmas, let us use them to derive the required inequality $(\star)$ and so complete the proof of the theorem. Lemma 2 is not directly applicable because, although the function $M(t)$ is continuous and nonnegative, it is not monotonic. In fact, $M(t)$ rises from $M(0) = \frac{3}{4}$ to $M(\frac{\pi}{12}) = \frac{\sqrt{3}}{2}$, then falls to $M(\frac{\pi}{6}) = \frac{3}{4}$ and $M(\frac{\pi}{3}) = 0$. The strategy is to write $M(t) = M_1(t) + M_2(t)$, where

$$M_1(t) = \begin{cases} \dfrac{3}{4}, & 0 \le t \le \dfrac{\pi}{6}, \\ M(t), & \dfrac{\pi}{6} \le t \le \dfrac{\pi}{2}. \end{cases}$$

Then $M_1(t)$ is nonincreasing, and Lemma 2 gives

$$\frac{4\pi}{9} \int_0^{\pi/2} M_1(t) J(t) \, dt \le \int_0^{\pi/2} t M_1(t) \, dt = \frac{\pi^2}{96} + \frac{\sqrt{3}\pi}{16} - \frac{3}{16}.$$

On the other hand, since $M(t) = M(\frac{\pi}{6} - t)$ for $0 \le t \le \frac{\pi}{6}$, a simple manipulation gives

$$\int_0^{\pi/2} M_2(t) J(t) \, dt = \frac{1}{2} \int_0^{\pi/6} M_2(t) \left[J(t) + J\left(\frac{\pi}{6} - t\right) \right] dt.$$

But $M_2(t) \ge 0$ by definition, and Lemma 1 gives

$$J(t) + J\left(\frac{\pi}{6} - t\right) \le \frac{3}{8} \qquad \text{for } 0 \le t \le \frac{\pi}{6}$$

(consider $6J(t) + 6J(\frac{\pi}{6} - t)$), so it follows that

$$\int_0^{\pi/2} M_2(t) J(t) \, dt \le \frac{3}{16} \int_0^{\pi/6} M_2(t) \, dt = \frac{3}{16} \left(\frac{\sqrt{3}}{4} - \frac{\pi}{8} \right).$$

Adding the two integral inequalities for M_1 and M_2, one arrives at the desired inequality $(\star)$. This completes the proof of the theorem. ∎

It now remains to establish the lemmas. The following elementary inequality will play a role.

Lemma 3. *Let $x_1, x_2, \ldots, x_n$ be positive numbers with sum no greater than π. Then $\sin^2 x_1 + \sin^2 x_2 + \cdots + \sin^2 x_n \le \frac{9}{4}$, with equality only for $n = 3$ and $x_1 = x_2 = x_3 = \frac{\pi}{3}$.*

Proof of Lemma 3. For $n \le 2$ the inequality holds trivially with the bound $2 < \frac{9}{4}$, so we may assume that $n \ge 3$. Let the function

$$F(x_1, x_2, \ldots, x_n) = \sin^2 x_1 + \sin^2 x_2 + \cdots + \sin^2 x_n$$

attain its maximum value at a point $(a_1, \ldots, a_n)$ with all $a_j \geq 0$ and $a_1 + \cdots + a_n = \pi$. In view of the simple inequality

$$\sin^2 a + \sin^2 b < \sin^2 (a + b), \qquad a > 0, b > 0, a + b < \frac{\pi}{2},$$

which is easily proved by expanding $\sin (a + b)$, each pair of nonzero coordinates a_i and a_j must have a sum $a_i + a_j \geq \frac{\pi}{2}$. Indeed, if $a_1 > 0, a_2 > 0$, and $a_1 + a_2 < \frac{\pi}{2}$, then

$$F(0, a_1 + a_2, a_3, \ldots, a_n) > F(a_1, a_2, a_3, \ldots, a_n),$$

contradicting the choice of $(a_1, \ldots, a_n)$ as a maximum point. Since $a_1 + \cdots + a_n = \pi$, it now follows that at most four of the a_j can be different from zero, so we may assume that $n \leq 4$. But if $n = 4$ and all $a_j \neq 0$, each pair of coordinates must have sum $\frac{\pi}{2}$ and so $a_j = \frac{\pi}{4}$ for $j = 1, 2, 3, 4$. But this is impossible, since

$$F\left(\frac{\pi}{4}, \frac{\pi}{4}, \frac{\pi}{4}, \frac{\pi}{4}\right) = 2 < \frac{9}{4} = F\left(0, \frac{\pi}{3}, \frac{\pi}{3}, \frac{\pi}{3}\right).$$

Therefore, we may assume that $n = 3$, and the proof is completed by straightforward calculus, using a Lagrange multiplier to accommodate the constraint $x_1 + x_2 + x_3 = \pi$. The critical points are easy to analyze when $n = 3$. ∎

Proof of Lemma 1. The idea is to shift the intervals of integration in the expressions for $J(t_k)$ so that Lemma 3 will apply to the sum of integrands. Define the linear functions $y_k(s)$ by $y_1(s) = s$, $y_2(s) = s + t_1 + t_2$, and

$$y_k(s) = s + t_1 + 2t_2 + \cdots + 2t_{k-1} + t_k, \qquad k = 3, \ldots, n.$$

Then $y_k(s) - t_k = y_{k-1}(s) + t_{k-1}, k = 2, \ldots, n$. Set

$$\alpha_k(s) = \theta(y_k(s) + t_k) - \theta(y_k(s) - t_k)$$

and observe that

$$J(t_k) = \frac{1}{2\pi} \int_0^{2\pi} \sin^2\left(\frac{\alpha_k(s)}{2}\right) ds, \qquad k = 1, \ldots, n.$$

But $\alpha_k(s) \geq 0$ and

$$\sum_{k=1}^{n} \alpha_k = \theta(y_1 + t_1) - \theta(y_1 - t_1) + \sum_{k=2}^{n} [\theta(y_k + t_k) - \theta(y_{k-1} + t_{k-1})]$$
$$= \theta(y_n + t_n) - \theta(y_1 - t_1) = 2\pi,$$

since $y_n + t_n - (y_1 - t_1) = 2(t_1 + \cdots + t_n) = 2\pi$ and $\theta(t + 2\pi) = \theta(t) + 2\pi$. Thus, Lemma 3 gives $\sum_{k=1}^n \sin^2 (\alpha_k(s)/2) \leq \frac{9}{4}$, and the desired inequality follows by integration. ∎

Proof of Lemma 2. Let $J^*(t) = J(t) - \frac{9t}{4\pi}$ and define the functional

$$I(P) = \int_0^{\pi/2} P(t)J^*(t)\,dt.$$

It is to be proved that $I(P) \leq 0$. Let

$$P_1(t) = \begin{cases} P(t) - P\left(\frac{\pi}{2} - t\right), & 0 \leq t \leq \frac{\pi}{4}, \\ 0, & \frac{\pi}{4} \leq t \leq \frac{\pi}{2}. \end{cases}$$

Then $0 \leq P_1(t) \leq P(t)$ and $P_1(t)$ is again nonincreasing. A bit of manipulation leads to the expression

$$I(P) = I(P_1) + \int_0^{\pi/4} P\left(\frac{\pi}{2} - t\right) \left[J^*(t) + J^*\left(\frac{\pi}{2} - t\right)\right] dt.$$

But Lemma 1 gives

$$2J^*(t) + 2J^*\left(\frac{\pi}{2} - t\right) = 2J(t) + 2J\left(\frac{\pi}{2} - t\right) - \frac{9}{4} \leq 0, \qquad 0 \leq t \leq \frac{\pi}{2},$$

showing that $I(P) \leq I(P_1)$. The process is now iterated. Let

$$P_2(t) = \begin{cases} P_1(t) - P_1\left(\frac{\pi}{4} - t\right), & 0 \leq t \leq \frac{\pi}{8}, \\ 0, & \frac{\pi}{8} \leq t \leq \frac{\pi}{2}. \end{cases}$$

and observe that $P_2(t)$ is nonincreasing, $0 \leq P_2(t) \leq P_1(t)$, and

$$I(P_1) = I(P_2) + \int_0^{\pi/8} P_1\left(\frac{\pi}{4} - t\right) \left[J^*(t) + J^*\left(\frac{\pi}{4} - t\right)\right] dt.$$

Lemma 1 now gives $4J^*(t) + 4J^*(\frac{\pi}{4} - t) \leq 0$, so $I(P_1) \leq I(P_2)$. Inductively defining

$$P_n(t) = \begin{cases} P_{n-1}(t) - P_{n-1}(\pi/2^n - t), & 0 \leq t \leq \pi/2^{n+1}, \\ 0, & \pi/2^{n+1} \leq t \leq \pi/2. \end{cases}$$

one finds in a similar way that $I(P_{n-1}) \leq I(P_n)$, whereas an inductive argument shows that $0 \leq P_n(t) \leq P_{n-1}(t)$. In particular, $0 \leq P_n(t) \leq P(t)$ for

$n = 1, 2, \ldots$. On the other hand, it is clear that $J(t) \to 0$ as $t \to 0$, so

$$I(P_n) = \int_0^{\pi/2^n} P_n(t) J^*(t) \, dt \to 0 \qquad \text{as } n \to \infty.$$

Since $I(P) \leq I(P_n)$, it follows that $I(P) \leq 0$. ∎

4.5. Coefficient Estimates

Variational methods are a well-established and powerful device for solving extremal problems in classes of analytic univalent functions. (An account may be found, for instance, in Duren [2], Chs. 9 and 10.) Corresponding variational techniques for general classes of harmonic mappings are not yet available, but for mappings onto *convex* domains the Radó–Kneser–Choquet theorem can be exploited to construct a calculus of variations that helps to solve extremal problems. Such a method was developed by Duren and Schober [1].

The method is most effective when specialized to harmonic self-mappings of the disk. According to the Radó–Kneser–Choquet theorem and its converse (Sections 3.2 and 3.3), a function f is a sense-preserving harmonic mapping of the unit disk $\mathbb{D}$ onto itself if and only if it has a Poisson integral representation

$$f(z) = \frac{1}{2\pi} \int_0^{2\pi} \frac{1 - |z|^2}{|e^{it} - z|^2} \, e^{i\theta(t)} \, dt,$$

where $e^{it} \mapsto e^{i\theta(t)}$ is a *weak homeomorphism* of the circle, a continuous sense-preserving mapping of the unit circle onto itself. This means that the function $t \mapsto \theta(t)$ is continuous and nondecreasing, mapping the interval $[0, 2\pi]$ onto an interval of length 2π, but it need not be *strictly* monotonic; it may have intervals of constancy.

For convenience, let $\mathcal{F}$ denote the family of sense-preserving harmonic mappings of $\mathbb{D}$ onto $\mathbb{D}$. Let ϕ be a continuous linear functional on $\mathcal{F}$ or, more generally, a continuous functional with a Fréchet differential. Continuity is with respect to the topology of local uniform convergence, so that $\phi(f_n) \to \phi(g)$ if a sequence $\{f_n\}$ of functions in $\mathcal{F}$ converges uniformly on compact subsets of $\mathbb{D}$ to a function $g \in \mathcal{F}$. Consider now the extremal problem of maximizing $\mathrm{Re}\,\{\phi(f)\}$ as f ranges over the family $\mathcal{F}$. Even if ϕ is bounded on $\mathcal{F}$, there may be no extremal function – the supremum of $\mathrm{Re}\,\{\phi\}$ need not be attained in $\mathcal{F}$. However, the closure of $\mathcal{F}$ consists of all functions f with a Poisson integral representation as displayed above, where the nondecreasing function $\theta(t)$ is allowed to have jump discontinuities. The corresponding function $e^{it} \mapsto e^{i\theta(t)}$ is called a *circle mapping*. Its Poisson integral will still be harmonic and univalent, except in degenerate cases where the range is a line segment or a point, but typically it will map the disk onto a polygonal region inscribed in the unit disk. We can therefore conclude that the supremum of Re$\{\phi\}$ for $f \in \mathscr{F}$ is attained by the Poisson integral of some circle mapping.

The corresponding extremal function $\theta(t)$ can now be subjected to a variation $\theta^*(t) = \theta(t) + \varepsilon\eta(t)$, where $e^{i\theta^*(t)}$ is a circle mapping with corresponding Poisson integral f^*. The extremal character of f then says that Re$\{\phi(f^*)\} \leq$ Re$\{\phi(f)\}$, and this information (with respect to all admissible variations) is typically enough to determine the function $\theta(t)$ and, hence, to determine the extremal function f. In practice, two types of variation are applied, one to an interval between jump points of $\theta(t)$, another to an interval in which $\theta(t)$ is not constant. The details are technical and will not be pursued here.

This method was applied successfully by Duren and Schober [1,2] to obtain the sharp bounds for the coefficients a_n and b_n of a function $f = h + \overline{g} \in \mathscr{F}$, where

$$h(z) = \sum_{n=0}^{\infty} a_n z^n, \qquad g(z) = \sum_{n=1}^{\infty} b_n z^n.$$

The first step is to obtain explicit formulas for the linear functionals a_n and b_n in terms of the function $\theta(t)$ that appears in the Poisson integral representation of f. Writing the Poisson kernel in the form

$$\frac{1 - |z|^2}{|e^{it} - z|^2} = \text{Re}\left\{\frac{e^{it} + z}{e^{it} - z}\right\} = \frac{1}{2}\left\{\frac{e^{it} + z}{e^{it} - z} + \frac{e^{-it} + \overline{z}}{e^{-it} - \overline{z}}\right\}$$

and expanding into geometric series, we see that

$$\frac{1 - |z|^2}{|e^{it} - z|^2} = 1 + 2\sum_{n=1}^{\infty} (e^{-int} z^n + e^{int} \overline{z}^n),$$

so that

$$a_n = \frac{1}{2\pi}\int_0^{2\pi} e^{i[\theta(t) - nt]}\, dt, \qquad n = 0, 1, 2, \ldots,$$

$$\overline{b_n} = \frac{1}{2\pi}\int_0^{2\pi} e^{i[\theta(t) + nt]}\, dt, \qquad n = 1, 2, \ldots.$$

Thus, the coefficient problems reduce to maximizing

$$\text{Re}\{a_n\} = \frac{1}{2\pi}\int_0^{2\pi} \cos\left[\theta(t) - nt\right] dt, \qquad n = 0, 1, 2, \ldots$$

and

$$\text{Re}\{b_n\} = \frac{1}{2\pi} \int_0^{2\pi} \cos[\theta(t) + nt]\,dt, \qquad n = 1, 2, \ldots$$

among all functions $\theta(t)$ that generate circle mappings.

Now if $\text{Re}\{a_n\}$ is maximized for some admissible function $\theta(t)$, we introduce a variation $\theta^* = \theta + \varepsilon\eta$ and use the inequality $\text{Re}\{a_n^*\} \le \text{Re}\{a_n\}$ to conclude from the Taylor expansion

$$\cos[\theta^*(t) - nt] = \cos[\theta(t) - nt] - \varepsilon\eta(t)\sin[\theta(t) - nt] + O(\varepsilon^2)$$

that

$$\int_0^{2\pi} \eta(t)\sin[\theta(t) - nt] \ge 0$$

for all admissible variations $\theta^* = \theta + \varepsilon\eta$ with $\varepsilon > 0$. With this information it is possible, by considering the explicit forms of the variations, to characterize the extremal functions $\theta(t)$. A similar treatment of the problem for b_n leads to the variational information

$$\int_0^{2\pi} \eta(t)\sin[\theta(t) + nt] \ge 0$$

for the extremal functions $\theta(t)$.

The two extremal problems, for a_n and b_n, are fundamentally different. First let us consider the problem of maximizing $\text{Re}\{a_n\}$. For $n = 0$ the sharp result is trivially $\text{Re}\{a_0\} < 1$ with unique extremal function $f(z) \equiv 1$, generated by $\theta(t) \equiv 0$ for $0 \le t < 2\pi$ and $\theta(2\pi) = 2\pi$. For $n = 1$ we can do no better than choose $\theta(t) = t$ for $0 \le t \le 2\pi$. Thus, $\text{Re}\{a_1\} \le 1$ and the only extremal function is the identity map $f(z) = z$. For $n \ge 2$ there are many extremal functions, but we can maximize $\text{Re}\{a_n\}$ for instance by choosing $\theta(t) = nt$ for $0 \le t \le 2\pi/n$ and $\theta(t) = 2\pi$ for $2\pi/n \le t \le 2\pi$. The sharp bound is

$$|a_n| \le \frac{1}{n}, \qquad n = 1, 2, \ldots.$$

The extremal functions f always belong to the family $\mathscr{F}$ and map the unit disk univalently onto itself. The associated functions $\theta(t)$ are piecewise linear, having intervals (mod 2π) of constancy whose lengths are integer multiples of $2\pi/n$ alternated with intervals of increase where $d\theta/dt = n$.

On the other hand, the coefficients b_n are found to have the sharp bounds

$$|b_n| < \frac{n+1}{n\pi} \sin\left(\frac{\pi}{n+1}\right), \qquad n = 1, 2, \ldots$$

for all $f \in \mathscr{F}$, where the "extremal functions" map the disk onto a regular $(n + 1)$-gon inscribed in the unit circle and are generated by circle mappings that are pure jump functions taking equally spaced values on the circle (the vertices of the polygon) on successive arcs of equal length. These are essentially the functions that were constructed in Section 4.2. For $n = 1$ the polygon degenerates to a line segment that forms a diameter of the circle, and the corresponding function f is not univalent.

The inequality $|a_n| \leq 1/n$ can also be proved in elementary fashion, without recourse to the variational method of Duren and Schober. The following argument is due to Sook Heui Jun [2]. Start with the formula

$$a_n = \frac{1}{2\pi} \int_0^{2\pi} e^{i\theta(t)} e^{-int} \, dt$$

and integrate by parts to obtain

$$2\pi a_n = \left[-\frac{1}{in} e^{-int} e^{i\theta(t)} \right]_0^{2\pi} + \frac{1}{n} \int_0^{2\pi} e^{-int} e^{i\theta(t)} \, d\theta(t).$$

But the first term vanishes by periodicity, so we have

$$2\pi n |a_n| = \left| \int_0^{2\pi} e^{-int} e^{i\theta(t)} \, d\theta(t) \right| \leq \int_0^{2\pi} d\theta(t) = 2\pi,$$

which implies that $|a_n| \leq 1/n$ for all sense-preserving harmonic mappings of the unit disk onto itself. By careful analysis of the cases of equality, the extremal functions can again be identified.

4.6. Schwarz's Lemma for Harmonic Mappings

The object of this section is to develop an appropriate analogue of the Schwarz lemma for complex-valued harmonic functions in the disk. The proof will make use of a special inequality for analytic functions (Lemma 2, below) which follows from the Schwarz lemma through the concept of subordination. The classical Schwarz lemma can be stated as follows.

Lemma 1 (Schwarz's Lemma). *Let f be analytic in the unit disk $\mathbb{D}$, with $f(0) = 0$ and $|f(z)| < 1$. Then $|f(z)| \leq |z|$, with strict inequality for all $z \neq 0$ in $\mathbb{D}$ unless f has the form $f(z) = \alpha z$ for some $\alpha \in \mathbb{C}$ with $|\alpha| = 1$. Also $|f'(0)| \leq 1$, with equality only for $f(z) = \alpha z$ with $|\alpha| = 1$.*

The Schwarz lemma applies naturally to the theory of subordination. A function f analytic in $\mathbb{D}$ is said to be *subordinate* to an analytic univalent function g (written $f \prec g$) if $f(0) = g(0)$ and $f(\mathbb{D}) \subset g(\mathbb{D})$. This implies that $\omega = g^{-1} \circ f$ is a *Schwarz function*: ω is analytic in $\mathbb{D}$ with $\omega(0) = 0$ and $|\omega(z)| < 1$. More generally, f is said to be subordinate to an analytic function g (not necessarily univalent) if $f = g \circ \omega$, where ω is a Schwarz function. This point of view is quite fruitful, as the following result indicates.

Lemma 2. *Let F be analytic in $\mathbb{D}$ and have the properties $F(0) = 0$ and $|\mathrm{Re}\,\{F(z)\}| < 1$. Then the inequalities*

$$|\mathrm{Re}\,\{F(z)\}| \le \frac{4}{\pi} \tan^{-1} |z|, \qquad |\mathrm{Im}\,\{F(z)\}| \le \frac{2}{\pi} \log \frac{1 + |z|}{1 - |z|}$$

hold and are sharp for each point z in $\mathbb{D}$.

It may be noted that $|z| < \frac{4}{\pi} \tan^{-1} |z|$ for $0 < |z| < 1$, so Schwarz's lemma gives a better bound under a stronger hypothesis.

Proof of Lemma 2. The function

$$G(z) = \frac{2i}{\pi} \log \frac{1 + z}{1 - z}$$

maps $\mathbb{D}$ conformally onto the vertical strip $|\mathrm{Re}\,\{w\}| < 1$. Thus, $F \prec G$ and $F = G \circ \omega$ for some Schwarz function ω. It follows that

$$\mathrm{Re}\,\{F(z)\} = -\frac{2}{\pi} \arg \left\{ \frac{1 + \omega(z)}{1 - \omega(z)} \right\}.$$

Observe now that the linear fractional mapping $w = \frac{1+z}{1-z}$ carries the circle $|z| = r < 1$ onto the circle $|w - w_0| = \rho$ with center $w_0 = (1 + r^2)/(1 - r^2)$ and radius $\rho = 2r/(1 - r^2)$. Therefore,

$$\left| \arg \left\{ \frac{1 + z}{1 - z} \right\} \right| \le \tan^{-1} \left(\frac{2r}{1 - r^2} \right) = 2 \tan^{-1} r,$$

and the Schwarz lemma gives

$$|\mathrm{Re}\,\{F(z)\}| \le \frac{4}{\pi} \tan^{-1} |z|,$$

with equality occurring only for $F(z) = G(\alpha z)$, where α is a suitable unimodular constant. The inequality for $\mathrm{Im}\,\{F(z)\}$ is proved similarly. ∎

It is now a short step to a harmonic version of the Schwarz lemma. The following sharp inequality is due to Heinz [2].

Theorem. *Let f be a complex-valued function harmonic in $\mathbb{D}$, with $f(0) = 0$ and $|f(z)| < 1$. Then $|f(z)| \leq \frac{4}{\pi} \tan^{-1} |z|$, and this inequality is sharp for each point z in $\mathbb{D}$. Furthermore, the bound is sharp everywhere (but is attained only at the origin) for* univalent *harmonic mappings f of $\mathbb{D}$ onto itself with $f(0) = 0$.*

Proof. For fixed θ, let F be the function analytic in $\mathbb{D}$ with $F(0) = 0$ and

$$\mathrm{Re}\,\{F(z)\} = \mathrm{Re}\,\{e^{-i\theta} f(z)\}.$$

Then $|\mathrm{Re}\{F(z)\}| < 1$, so Lemma 2 shows that

$$|\mathrm{Re}\,\{e^{-i\theta} f(z)\}| \leq \frac{4}{\pi} \tan^{-1} |z|.$$

But this inequality holds for all θ, so it follows that

$$|f(z)| \leq \frac{4}{\pi} \tan^{-1} |z|.$$

An analysis of the proof reveals that equality can occur only for functions of the form

$$f(z) = \frac{2\beta}{\pi} \arg \left\{ \frac{1 + \alpha z}{1 - \alpha z} \right\}, \qquad |\alpha| = |\beta| = 1,$$

whose values are confined to a diametral segment of $\mathbb{D}$. Up to rotation, these are precisely the Poisson integrals of $e^{i\theta}(t)$, where

$$\theta(t) = \begin{cases} 0, & 0 \leq t \leq \pi \\ \pi, & \pi < t < 2\pi. \end{cases}$$

Approximating this step function by a continuous increasing function $\theta(t)$ with $\theta(0) = 0$, $\theta(2\pi) = 2\pi$, and

$$\int_0^{2\pi} e^{i\theta(t)} \, dt = 0,$$

one can produce by a Poisson integral (see Section 4.1) a univalent harmonic mapping of $\mathbb{D}$ onto $\mathbb{D}$ with $f(0) = 0$ and $|f(z)|$ arbitrarily close to $\frac{4}{\pi} \tan^{-1} |z|$, where z is a prescribed point of the disk. The bound is therefore sharp even for harmonic mappings of the disk onto itself. ∎

5

Harmonic Univalent Functions

5.1. Normalizations

The object of this chapter is to study univalent harmonic functions as generalizations of univalent analytic functions. The point of departure is the canonical representation

$$f = h + \overline{g}, \qquad g(0) = 0,$$

of a harmonic function f in the unit disk $\mathbb{D}$ as the sum of an analytic function h and the conjugate of an analytic function g. With the convention that $g(0) = 0$, the representation is unique (see Section 1.2). The power series expansions of h and g are denoted by

$$h(z) = \sum_{n=0}^{\infty} a_n z^n \qquad \text{and} \qquad g(z) = \sum_{n=1}^{\infty} b_n z^n.$$

If f is a sense-preserving harmonic mapping of $\mathbb{D}$ onto some other region, then by Lewy's theorem (Section 2.2) its Jacobian is strictly positive. Equivalently, the inequality $|g'(z)| < |h'(z)|$ holds for all $z \in \mathbb{D}$. This shows in particular that $h'(z) \neq 0$, so there is no loss of generality in supposing that $h(0) = 0$ and $h'(0) = 1$. The class of all sense-preserving harmonic mappings of the disk with $a_0 = b_0 = 0$ and $a_1 = 1$ will be denoted by S_H. Thus, S_H contains the standard class S of analytic univalent functions. Although the analytic part h of a function $f \in S_H$ is locally univalent, it will become apparent that it need not be univalent.

We shall see in the next section that S_H is a *normal family*: every sequence of functions in S_H has a subsequence that converges locally uniformly in $\mathbb{D}$. On the other hand, S_H is not a *compact family*; it is not preserved under passage to locally uniform limits. The limit function is necessarily harmonic in $\mathbb{D}$, but it need not be univalent. To see this, simply consider the sequence of affine mappings $f_n \in S_H$ defined by

$$f_n(z) = z + \frac{n}{n+1}\overline{z}.$$

Then $f_n(z) \to f(z) = 2x$ (where $z = x + iy$) locally uniformly in $\mathbb{D}$, but f is not univalent (nor is it constant).

There is one further normalization, however, which succeeds in producing a compact normal family. The idea is simply to follow a given harmonic mapping by a suitable affine mapping to make $b_1 = 0$. Specifically, each $f \in S_H$ has the property $|b_1| < |a_1| = 1$, and so the function

$$\varphi(w) = \frac{w - \overline{b_1}\,\overline{w}}{1 - |b_1|^2}$$

is a sense-preserving affine mapping. The composition

$$f_0 = \varphi \circ f = h_0 + \overline{g_0}$$

is therefore a sense-preserving harmonic mapping that is easily seen to have the properties

$$h_0(0) = g_0(0) = 0, \qquad h_0'(0) = 1, \qquad \text{and} \qquad g_0'(0) = 0.$$

Thus, $f_0 \in S_H$, and it has the additional property that $g_0'(0) = 0$. The class of functions $f \in S_H$ with $g'(0) = 0$ will be denoted by S_H^0.

If $f = h + \overline{g}$ is in S_H^0, then $g'(0) = 0$ and $|g'(z)/h'(z)| < 1$, so it follows from the classical Schwarz lemma that $|g'(z)| \le |z||h'(z)|$. In terms of the dilatation $\omega = \overline{f_{\overline{z}}}/f_z$, this says that $|\omega(z)| \le |z|$ if $f \in S_H^0$.

In the next section it will be shown that S_H^0 is a compact normal family. This property makes S_H^0 seem more promising than S_H as a "correct" generalization of the family S of analytic univalent functions.

The mapping $f \longmapsto f_0 = \varphi \circ f$, which carries $f \in S_H$ to $f_0 \in S_H^0$, is invertible, with

$$f = f_0 + \overline{b_1 f_0}.$$

Thus, for each specified value of b_1 with $|b_1| < 1$ there is a unique function $f \in S_H$ with $g'(0) = b_1$ that corresponds to a given function $f_0 \in S_H^0$ under the standard affine mapping φ.

5.2. Normal Families

In the terminology of the preceding section, we can now state a theorem that has significant implications in the study of extremal problems for harmonic mappings.

Theorem. *The family S_H is normal. The family S_H^0 is normal and compact.*

This result is due to Clunie and Sheil-Small [1], and their proof will be adapted here. An immediate corollary is the existence of an extremal function in S_H^0 for any problem of maximizing a (real-valued) functional that is continuous with respect to the topology of local uniform convergence.

The proof of the theorem is not easy. It will be based on the observation that Montel's criterion for the normality of families of analytic functions is equally valid for families of harmonic functions. Recall (*cf.* Section 1.5) that a family $\mathscr{F}$ of functions defined on a common domain Ω is said to be *locally bounded* if the functions in $\mathscr{F}$ are uniformly bounded in some neighborhood of each point of Ω. More precisely, to each point z_0 in Ω there corresponds a number $M > 0$ and a neighborhood V of z_0 such that $|f(z)| \leq M$ for each $f \in \mathscr{F}$ and for all points $z \in V$. Equivalently, the functions in $\mathscr{F}$ are uniformly bounded on each compact subset of Ω. Montel's theorem says that a family of analytic functions is normal if (and only if) it is locally bounded. The proof is essentially the same when the principle is extended to families of harmonic functions. If such a family $\mathscr{F}$ is locally bounded, then by the Poisson formula the derived family

$$\mathscr{F}' = \left\{ \frac{\partial f}{\partial x} : f \in \mathscr{F} \right\} \cup \left\{ \frac{\partial f}{\partial y} : f \in \mathscr{F} \right\}$$

is also locally bounded. This implies that $\mathscr{F}$ is equicontinuous when restricted to an arbitrary compact subset of Ω. Thus, the normality of $\mathscr{F}$ follows from the Arzela–Ascoli theorem and a diagonalization argument. (For further details see Ahlfors [3], p. 219, or Duren [2], p. 7.)

To prove the normality of S_H, it therefore suffices to establish its local boundedness. With the notation

$$M_\infty(r, f) = \max_{|z|=r} |f(z)|,$$

it follows from the relation $f = f_0 + \overline{b_1} \, \overline{f_0}$ of the previous section that

$$M_\infty(r, f) \leq 2M_\infty(r, f_0).$$

Thus, it is enough to show that S_H^0 is locally bounded. The proof relies on a form of the Schwarz lemma for quasiconformal mappings, stated as follows.

Lemma. *Let G be a K-quasiconformal mapping of the unit disk into itself with $G(0) = 0$. Then $|G(z)| \leq \varphi_K(|z|)$, where*

$$\varphi_K(r) = \mu^{-1}\left(\frac{\mu(r)}{K} \right)$$

*and $\mu(r)$ is the module of the ring domain bounded by the unit circle and the
radial segment from 0 to r. For each fixed $K > 1$, the bound $\varphi_K(r)$ increases
with r from 0 to 1.*

The *module* of a ring domain (*i.e.*, a doubly connected domain) is $\mu =
1/2\pi \log M$, where $M = r_2/r_1$ is the modulus as defined in Section 8.1. Here
we are dealing with the module of a *Grötzsch ring*, as it is called, and $\mu(r)$
can be expressed in terms of the elliptic integral

$$\mathscr{K}(k) = \int_0^{\pi/2} \{1 - k^2 \sin^2 \theta\}^{-1/2} \, d\theta, \qquad 0 < k < 1.$$

With the help of a Schwarz–Christoffel mapping of the upper half-plane onto
a rectangle, it is found (*cf.* Nehari [2], p. 293) that

$$\mu(r) = \frac{1}{4} \frac{\mathscr{K}(r')}{\mathscr{K}(r)}, \qquad r' = \sqrt{1 - r^2}.$$

However, for present purposes the specific formula is unimportant. All that
matters is the existence of a locally uniform bound less than 1 .

The lemma is due to Hersch and Pfluger [1]. For a proof, the reader is
referred to the literature on quasiconformal mappings (see, for instance, Lehto
and Virtanen [1], p. 63).

Proof of Theorem. Let $f = h + \overline{g} \in S_H^0$. Then by the classical Schwarz
lemma, $|g'(z)| \le |z||h'(z)|$. (See Section 5.1.) Thus, for each fixed R with
$0 < R < 1$, the mapping $w = f(Rz)$ is K-quasiconformal in $\mathbb{D}$ with $K =
(1 + R)/(1 - R)$. (See Section 1.2 for a brief discussion of quasiconformal
mappings.) Now define the *conformal associate* of f to be the conformal
mapping F of $\mathbb{D}$ onto the range $f(\mathbb{D})$, normalized by the conditions $F(0) = 0$
and $F'(0) > 0$. Let $G(z) = F^{-1}(f(Rz))$. Then G is again K-quasiconformal
(see Section 1.2), and it maps the disk into itself with $G(0) = 0$. By the lemma,
$|G(z)| \le \varphi_K(|z|)$ for all $z \in D$. This inequality may be rewritten as

$$|F^{-1}(f(z))| \le \varphi_K\left(\frac{|z|}{R}\right), \qquad |z| \le R = \frac{K - 1}{K + 1}.$$

On the other hand, the growth theorem for univalent analytic functions (see
Section 1.5) says that

$$|F(z)| \le \frac{|F'(0)||z|}{(1 - |z|)^2}, \qquad |z| < 1.$$

It therefore follows that

$$|f(z)| = |F(F^{-1}(f(z)))| \le \frac{|F'(0)|\varphi_K(r/R)}{[1 - \varphi_K(r/R)]^2}, \qquad |z| \le r < R.$$

In order to estimate $F'(0)$, we invoke the Koebe one-quarter theorem (see Section 1.5) in the form $|F'(0)| \leq 4|w_0|$, where w_0 is any value omitted by F. But it was shown in the proof of Radó's theorem (Section 2.4) that

$$cR^2 \leq |f_z(0)|^2 + |f_{\bar{z}}(0)|^2$$

for any harmonic mapping of the unit disk whose range contains the disk $|w| < R$, where c is a positive absolute constant. Since $f_z(0) = 1$ and $f_{\bar{z}}(0) = 0$ for every function $f \in S_H^0$, this shows that f omits a value of modulus $|w_0| = 1/\sqrt{c}$. It follows that $|F'(0)| \leq 4/\sqrt{c}$, completing the proof that S_H^0 is locally bounded and therefore normal. As remarked earlier, it follows that S_H is normal.

It remains to show that S_H^0 is compact. For this purpose, suppose that $f_n = h_n + \overline{g_n} \in S_H^0$ and that $f_n \to f$ uniformly on compact subsets of $\mathbb{D}$. Then f is harmonic, and so it has a canonical representation $f = h + \overline{g}$. It is easy to see that $h_n \to h$ and $g_n \to g$ locally uniformly, and that $g'(0) = 0$ and $h'(0) = 1$. In particular, h is not constant and so by Hurwitz's theorem $h'(z) \neq 0$ in $\mathbb{D}$. However, the Schwarz lemma gives $|g'(z)| \leq |z||h'(z)|$, and the Jacobian

$$J_f(z) = |h'(z)|^2 - |g'(z)|^2 \geq (1 - |z|^2)|h'(z)|^2 > 0$$

throughout the disk. In other words, f is locally univalent in $\mathbb{D}$. Because it is also the uniform limit of univalent functions on compact subsets of $\mathbb{D}$, it follows from the argument principle (see Section 1.3) that f is univalent in $\mathbb{D}$. Hence, the limit function f again belongs to S_H^0, and we have shown that S_H^0 is a compact family. This concludes the proof. ∎

5.3. The Harmonic Koebe Function

The classical (analytic) Koebe function

$$k(z) = z(1 - z)^{-2} = z + 2z^2 + 3z^3 + \cdots$$

maps the unit disk conformally onto the complex plane deprived of a half-line along the negative real axis from $-\frac{1}{4}$ to infinity. It has long been known to play the role of extremal function for many extremal problems over the class S of analytic univalent functions (see Section 1.5). The harmonic Koebe function, now to be constructed, is the probable analogue of the Koebe function for the class S_H^0 of harmonic univalent functions. It will be seen to map the disk harmonically onto the full plane minus the part of the negative real axis from $-\frac{1}{6}$ to infinity. The extremal elements in its construction suggest that it plays a role in S_H^0 similar to that of the Koebe function in S, but this has not yet been fully confirmed.

The construction of the harmonic Koebe function is due to Clunie and Sheil-Small [1] and is based again on their general result (Theorem 1 in Section 3.4) about mappings onto domains convex in the horizontal direction (CHD). According to that theorem, a locally univalent harmonic function $f = h + \bar{g}$ is univalent and its range is CHD if and only if the analytic function $h - g$ has the same property.

Observe now that the Koebe function $k(z) = z(1 - z)^{-2}$ is univalent in $\mathbb{D}$ and its range is convex in the horizontal direction. In view of the theorem, a locally univalent harmonic mapping $f = h + \bar{g}$ will have the same properties if $h - g = k$. As in Section 3.4, the local univalence of f is assured by requiring that $g'(z) = zh'(z)$. In other words, the dilatation of f is prescribed to be $\omega(z) = z$. After differentiation, the two conditions reduce to the pair of linear equations

$$h'(z) - g'(z) = k'(z)$$
$$zh'(z) - g'(z) = 0.$$

Noting that $k'(z) = (1 + z)(1 - z)^{-3}$, one is therefore led to the formulas

$$h'(z) = \frac{(1 + z)}{(1 - z)^4}, \qquad g'(z) = \frac{z(1 + z)}{(1 - z)^4}.$$

Integration now gives

$$h(z) = \frac{z - \frac{1}{2}z^2 + \frac{1}{6}z^3}{(1 - z)^3}, \qquad g(z) = \frac{\frac{1}{2}z^2 + \frac{1}{6}z^3}{(1 - z)^3},$$

under the assumption that $h(0) = g(0) = 0$.

The resulting function $K = h + \bar{g}$, with h and g as displayed above, is known as the *harmonic Koebe function*. It is clear from the construction that K is univalent and sense-preserving in the disk and, in fact, that $K \in S_H^0$. It is also clear that the range of K is convex in the horizontal direction and is symmetric with respect to the real axis. What is not so clear is the actual range of K.

In a first attempt to determine the range of K, one might try to find the image of the unit circle. The explicit formulas show that K is well-behaved at each boundary point except for $z = 1$. In fact, K has a harmonic extension to $\mathbb{C} \setminus \{1\}$. Calculations reveal, however, that $K(e^{it}) = -\frac{1}{6}$ for every point $e^{it} \neq 1$ on the circle! To see this economically, write

$$K = h + \bar{g} = (h - g) + 2\,\mathrm{Re}\,\{g\} = k + 2\,\mathrm{Re}\,\{g\},$$

or

$$K(z) = \text{Re}\left\{\frac{z + \frac{1}{3}z^3}{(1-z)^3}\right\} + i\,\text{Im}\left\{\frac{z}{(1-z)^2}\right\}.$$

Since the analytic Koebe function k maps the disk onto the plane with a slit along part of the real axis, it is clear geometrically that $\text{Im}\{k(z)\} = 0$ on the unit circle $|z| = 1, z \neq 1$. A straightforward but tedious calculation shows that

$$\text{Re}\left\{\frac{z + \frac{1}{3}z^3}{(1-z)^3}\right\} \equiv -\frac{1}{6} \qquad \text{for } |z| = 1, z \neq 1,$$

thus verifying that $K(z) \equiv -\frac{1}{6}$ on the unit circle except at the point $z = 1$.

It is more instructive, however, to interpose the transformation

$$\zeta = \frac{1+z}{1-z} = \xi + i\eta,$$

which maps $\mathbb{D}$ onto the right half-plane $\text{Re}\{\zeta\} > 0$. Calculations show that

$$K(z) = \text{Re}\left\{\frac{1}{6}(\zeta^3 - 1)\right\} + i\,\text{Im}\left\{\frac{1}{4}(\zeta^2 - 1)\right\}$$

$$= \frac{1}{6}(\xi^3 - 3\xi\eta^2 - 1) + i\frac{1}{2}\xi\eta, \qquad \xi > 0.$$

Observe now that each point z on the unit circle $(z \neq 1)$ is carried onto a point ζ on the imaginary axis so that $\xi = 0$ and $K(z) = -\frac{1}{6}$. Next observe that the positive real axis

$$\{\zeta = \xi + i\eta : \xi > 0, \eta = 0\}$$

is mapped monotonically to the real interval $(-\frac{1}{6}, \infty)$. Finally, each hyperbola $\xi\eta = c$, where $c \neq 0$ is a real constant, is carried univalently to the set

$$\{w = u + i\frac{c}{2} : u = \frac{1}{6}(\xi^3 - 3c^2\xi^{-1} - 1), \xi > 0\},$$

which is the entire line $\{w = u + i\frac{c}{2} : -\infty < u < \infty\}$. This proves directly that K is univalent in the disk and maps it onto the entire plane minus the real interval $(-\infty, -\frac{1}{6}]$. Figure 5.1 illustrates the argument. The action of K is depicted in Figure 5.2.

It still remains to be explained how K can carry the whole unit circle, except for $z = 1$, to the single point $w = -\frac{1}{6}$, yet have an image whose boundary is the whole interval $(-\infty, -\frac{1}{6}]$. It is fairly obvious that the explanation must lie in the behavior of K near the point 1. But as z tends to 1 nontangentially, $K(z)$ tends to infinity. To see this, consider the behavior of $K(z)$ on the

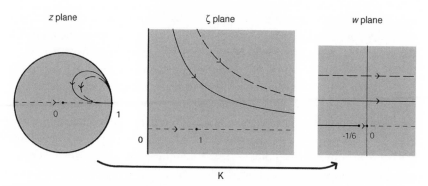

Figure 5.1. Range of harmonic Koebe function.

curve corresponding to the ray $\eta = c\xi$, $\xi > 0$. As $\xi \to +\infty$, it is evident that $K(z) \to \infty$. However, as $z \to 1$ along the *tangential* curve corresponding to the curve $\eta = c/\sqrt{\xi}$ ($\xi > 0$) in the ζ plane, where $c \neq 0$ is a real constant, one finds that

$$K(z) = \frac{1}{6}(\xi^3 - 3c^2 - 1) + i\frac{c}{2}\sqrt{\xi} \to -\frac{1}{6}(3c^2 + 1)$$

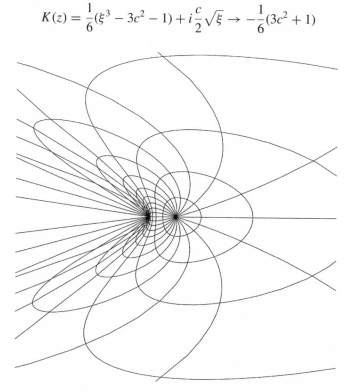

Figure 5.2. The harmonic Koebe function.

as $\xi \to 0$. As c ranges over the real line, this accounts for all of the "missing" boundary points. Notice that, if $c > 0$, then the corresponding boundary point is approached from above; whereas if $c < 0$, it is approached from below.

The exotic boundary behavior exhibited by the harmonic Koebe function, the mapping of boundary arcs to points, is actually rather typical of harmonic mappings produced by the shear construction. Several other examples were found in Section 3.4. For instance, the harmonic half-plane mapping $L = \text{Re}\,\{\ell\} + i\,\text{Im}\,\{k\}$ maps the disk onto the half-plane $\text{Re}\,\{w\} > -\frac{1}{2}$ but sends the entire boundary to the point $-\frac{1}{2}$, except for the point $z = 1$, which goes to infinity. This kind of behavior, where the boundary function is constant on an arc, can occur only when the dilatation has unit modulus on the corresponding boundary arc. The phenomenon will be discussed further in Sections 7.3 and 7.4.

5.4. Coefficient Conjectures

With slight change of notation, the harmonic Koebe function can be written as $K = H + \overline{G}$, where H and G are the rational functions constructed in the preceding section:

$$H(z) = \frac{z - \frac{1}{2}z^2 + \frac{1}{6}z^3}{(1-z)^3}, \qquad G(z) = \frac{\frac{1}{2}z^2 + \frac{1}{6}z^3}{(1-z)^3}.$$

In the notation

$$H(z) = \sum_{n=1}^{\infty} A_n z^n, \qquad G(z) = \sum_{n=2}^{\infty} B_n z^n,$$

the coefficients of H and G are found to be

$$A_n = \frac{1}{6}(2n+1)(n+1), \qquad B_n = \frac{1}{6}(2n-1)(n-1).$$

These calculations suggest the conjectures

$$|a_n| \le \frac{1}{6}(2n+1)(n+1), \qquad |b_n| \le \frac{1}{6}(2n-1)(n-1)$$

for all functions f in S_H^0 and for all indices $n \ge 2$. Clunie and Sheil-Small [1] proved that these bounds hold for functions in S_H^0 with real coefficients and, more generally, for typically real functions (see Section 6.6). Sheil-Small [3] established the same estimates for starlike mappings (see Section 6.7). In the full class S_H^0, however, only the elementary inequality $|b_2| \le \frac{1}{2}$ has been verified so far.

Theorem. *For all functions $f \in S_H^0$, the sharp inequality $|b_2| \le \frac{1}{2}$ holds.*

Proof. Since $f = h + \overline{g}$ is sense-preserving, its analytic dilatation $\omega = g'/h'$ satisfies $|\omega(z)| < 1$. Also, $f \in S_H^0$ means that $g'(0) = 0$, so $\omega(0) = 0$. Hence, the Schwarz lemma gives $|\omega'(0)| \le 1$, with equality if and only if $\omega(z) = e^{i\alpha}z$. But

$$\omega(z) = \frac{g'(z)}{h'(z)} = \frac{2b_2 z + 3b_3 z^2 + \cdots}{1 + 2a_2 z + \cdots}$$

$$= 2b_2 z + (3b_3 - 4a_2 b_2)z^2 + \cdots,$$

so the conclusion is that $|2b_2| \le 1$, with equality holding if and only if the analytic dilatation ω is a rotation of the identity. ∎

Since $b_2 = \frac{1}{2}$ whenever $\omega(z) = z$, it is all too clear that the inequality $|b_2| \le \frac{1}{2}$ admits an abundance of extremal functions. For instance, the harmonic Koebe function and the simple mapping $f(z) = z + \frac{1}{2}\overline{z}^2$ both have $b_2 = \frac{1}{2}$. The construction of the harmonic Koebe function can be modified to produce many other harmonic mappings (with range convex in the direction of the real axis) having $b_2 = \frac{1}{2}$.

The conjecture $|a_2| \le \frac{5}{2}$ is of special importance. As Sheil-Small [3] has pointed out, its truth would imply the sharp forms of various distortion and covering theorems for the class S_H^0. At present only very crude estimates of $|a_2|$ are available. Clunie and Sheil-Small [1] showed that $|a_2| < 12, 172$, and Sheil-Small [3] improved that estimate to $|a_2| < 57$. A small modification of his argument gives $|a_2| < 49$, as will be shown later (Section 6.3), but there is still plenty of room for improvement.

Another attractive conjecture is that

$$||a_n| - |b_n|| \le n, \qquad n = 2, 3, \ldots,$$

a generalization of the Bieberbach conjecture to the class S_H^0, suggested by the identity $A_n - B_n = n$ for the coefficients of the harmonic Koebe function. Clunie and Sheil-Small [1] verified this conjecture for typically real functions, and Sheil-Small [3] proved it for starlike functions (see Sections 6.6 and 6.7), but it remains open for the full class S_H^0. The inequality $||a_2| - |b_2|| \le 2$ would imply $|a_2| \le \frac{5}{2}$, since $|b_2| \le \frac{1}{2}$.

Because each function $f \in S_H$ has the form $f = f_0 + \overline{b_1}\,\overline{f_0}$ for some $f_0 \in S_H^0$ and $|b_1| < 1$, it is easy to see that the inequalities

$$|a_n| \le \frac{1}{6}(2n + 1)(n + 1), \qquad |b_n| \le \frac{1}{6}(2n - 1)(n - 1)$$

for functions $f_0 \in S_H^0$ would imply the sharp bounds (not attained)

$$|a_n| < \frac{1}{3}(2n^2 + 1), \qquad |b_n| < \frac{1}{3}(2n^2 + 1)$$

for functions $f \in S_H$. In particular, the inequality $|a_2| \leq \frac{5}{2}$ in S_H^0 would imply that $|a_2| < 3$ for all functions of class S_H.

6

Extremal Problems

6.1. Minimum Area

This chapter presents solutions and partial solutions to some extremal problems in the class S_H^0. We begin with a relatively simple question that can be answered completely. Among all functions in S_H^0, which ones map the disk to a region of smallest area?

It is not at all obvious *a priori* that a minimum is actually attained for some function in S_H^0, even though it is a compact normal family. In fact, the corresponding problem for the larger class S_H has no extremal functions. To see this, consider the affine mappings $f(z) = z + \beta \bar{z}, 0 < |\beta| < 1$, which belong to S_H but not to S_H^0. Here the Jacobian is $J_f(z) \equiv 1 - |\beta|^2$, and so the image $f(\mathbb{D})$ has area $A = \pi(1 - |\beta|^2)$, which can be made arbitrarily small. Thus, S_H contains no area-minimizing functions.

Nevertheless, the class S_H^0 does contain area-minimizing functions, and they can all be found by direct estimation. Each function in S_H^0 can be written as $f = h + \bar{g}$ for some analytic functions h and g of the forms

$$h(z) = z + \sum_{n=2}^{\infty} a_n z^n, \qquad g(z) = \sum_{n=2}^{\infty} b_n z^n.$$

Then $g' = \omega h'$, where $|\omega(z)| < 1$ and $\omega(0) = 0$; hence, by the Schwarz lemma, $|\omega(z)| \leq |z|$. The Jacobian of f is $|h'|^2 - |g'|^2$, and so the area A of $f(\mathbb{D})$ is

$$A = \iint_{\mathbb{D}} |h'(z)|^2 - |g'(z)|^2) \, dx \, dy$$

$$= \iint_{\mathbb{D}} \left(1 - |\omega(z)|^2\right) |h'(z)|^2 \, dx \, dy \geq \iint_{\mathbb{D}} (1 - |z|^2)|h'(z)|^2 \, dx \, dy$$

$$= \pi \sum_{n=1}^{\infty} n|a_n|^2 - \iint_{\mathbb{D}} \left| \sum_{n=1}^{\infty} n a_n z^n \right|^2 \, dx \, dy$$

$$= \pi \sum_{n=1}^{\infty} n \left(1 - \frac{n}{n+1}\right) |a_n|^2 = \frac{\pi}{2} + \pi \sum_{n=2}^{\infty} n \left(1 - \frac{n}{n+1}\right) |a_n|^2,$$

as shown by a standard calculation in polar coordinates. The last sum is clearly minimized by choosing $a_n = 0$ for all $n \geq 2$. The minimum area is attained only if $|\omega(z)| \equiv |z|$, or $\omega(z) = e^{i\alpha} z$. Hence, $g'(z) = e^{i\alpha} z h'(z) = e^{i\alpha} z$, which implies $|b_2| = \frac{1}{2}$ and $b_n = 0$ for all $n \geq 3$. But the function $f(z) = z + \frac{1}{2}\bar{z}^2$ is known to be univalent (see Section 1.1), so the problem is solved.

Theorem. *The area of the image of each function f in S_H^0 is greater than or equal to $\frac{\pi}{2}$, and this is a minimum attained only by the function $f(z) = z + \frac{1}{2}\bar{z}^2$ and its rotations.*

6.2. Covering Theorems

Our next objective is to show that some well-known covering theorems for analytic univalent functions have partial extensions to harmonic mappings. According to the Koebe one-quarter theorem, the range of every function of class S contains the open disk $|w| < 1/4$. The Koebe function $k(z) = z(1 - z)^{-2}$, which maps the unit disk onto the whole plane minus the half-line from $-\frac{1}{4}$ to infinity, shows that the radius $\frac{1}{4}$ is best possible. The classical theory of analytic univalent functions was outlined in Section 1.5.

The harmonic Koebe function, as constructed in Section 5.3, maps the disk onto the plane minus the half-line from $-\frac{1}{6}$ to infinity. It suggests that perhaps each function in S_H^0 will cover the disk $|w| < 1/6$. The answer is unknown, but we shall present here a proof by Clunie and Sheil-Small [1] that at least the disk $|w| < 1/16$ is always covered. We begin, however, with a result in the opposite direction, showing that a normalized harmonic mapping cannot cover too large a disk.

Theorem 1. *Each function in S_H omits some point on the circle $|w| = \frac{2\pi\sqrt{6}}{9} = 1.710\ldots$. Each function in S_H^0 omits some point on the circle $|w| = \frac{2\pi\sqrt{3}}{9} = 1.209\ldots$, but need not omit any point of smaller modulus.*

The corresponding result for analytic mappings is that each function in S must omit a point on the unit circle; it cannot cover a disk larger than $\mathbb{D}$ itself. In fact, the Schwarz lemma shows that the only function $f \in S$ for which $f(\mathbb{D}) \supset \mathbb{D}$ is the identity. Specifically, if g is the restriction of f^{-1} to $\mathbb{D}$, then $g(\mathbb{D}) \subset \mathbb{D}$, $g(0) = 0$, and $g'(0) = 1$. Hence, $g(w) \equiv w$ by the Schwarz lemma.

Proof of Theorem. In the proof of Radó's theorem (Section 2.4), it was shown that if a harmonic mapping f has the property $f(0) = 0$ and contains a disk $|w| < R$ in its range, then

$$cR^2 \leq |f_z(0)|^2 + |f_{\bar{z}}(0)|^2 = |a_1|^2 + |b_1|^2,$$

where c is the Heinz constant. But in Section 4.4 the sharp value of the Heinz constant was found to be $c = 27/4\pi^2$. Thus,

$$R \leq \frac{2\pi\sqrt{3}}{9} \left(|a_1|^2 + |b_1|^2\right)^{1/2}.$$

If $f \in S_H$, then $a_1 = 1$ and $|b_1| \leq 1$, so this implies $R \leq 2\pi\sqrt{6}/9$. If $f \in S_H^0$, then $a_1 = 1$ and $b_1 = 0$, so it implies $R \leq 2\pi\sqrt{3}/9$. In either case it follows that some omitted value must lie on the circle of given radius.

To see that the bound $2\pi\sqrt{3}/9$ is best possible, let

$$f(z) = \frac{1}{\pi} \sum_{k=0}^{2} \alpha^k \arg \left\{ \frac{z - \beta^{2k+1}}{z - \beta^{2k-1}} \right\}, \qquad \beta = e^{i\pi/3}, \qquad \alpha = \beta^2,$$

be the canonical mapping of $\mathbb{D}$ onto the interior of an equilateral triangle, as constructed in Section 4.2. Then $f_z(0) = \frac{3\sqrt{3}}{2\pi}$ and $f_{\bar{z}}(0) = 0$, as observed in Section 4.4. Consider now a sequence $\{f_n\}$ of sense-preserving harmonic mappings of $\mathbb{D}$ onto itself, converging locally uniformly to f, constructed so that $(\partial f_n/\partial \bar{z})(0) = 0$. Such functions can be obtained by rotating the approximations specified in Section 4.1. They have the symmetry property $f_n(\alpha z) = \alpha f_n(z)$, which implies that

$$\overline{\alpha}(\partial f_n/\partial \bar{z})(0) = \alpha(\partial f_n/\partial \bar{z})(0),$$

so that $(\partial f_n/\partial \bar{z})(0) = 0$. Now let $r_n = 1/(\partial f_n/\partial z)(0)$, noting that $r_n > 0$, and define $g_n = r_n f_n$. Then $g_n \in S_H^0$ and it maps $\mathbb{D}$ onto the disk $|w| < r_n$. However, r_n converges to $1/(\partial f/\partial z)(0) = 2\pi\sqrt{3}/9$ as n tends to infinity. Thus, a function in S_H^0 need not omit any value of prescribed modulus smaller than $2\pi\sqrt{3}/9$. ∎

The estimates of the theorem are due to Clunie and Sheil-Small [1]. Richard Laugesen (private communication) pointed out the preceding construction showing that the radius $2\pi\sqrt{3}/9$ is best possible for functions in S_H^0. On the other hand, Richard Hall [3] showed that for functions in S_H the radius $2\pi\sqrt{6}/9$ can be decreased to $\frac{\pi}{2} = 1.570\ldots$, which is best possible. In fact, by methods similar to his proof of the Heinz conjecture (see Section 4.4), Hall proved a conjecture of Ullman and Titus [1] to the effect that $|a_1| \geq \frac{2}{\pi}$ for sense-preserving harmonic mappings of the unit disk onto itself that fix the origin. Thus, the range of a function $f \in S_H$ cannot contain any disk $|w| < R$ with $R > \frac{\pi}{2}$, so it must omit a point on the circle $|w| = \frac{\pi}{2}$. To see that $\frac{\pi}{2}$ is the best possible, we need only construct (as in Section 4.1) a sequence of harmonic mappings of $\mathbb{D}$ onto $\mathbb{D}$, fixing the origin and converging locally uniformly to the function

$$f(z) = \frac{2}{\pi} \arg \left\{ \frac{1+z}{1-z} \right\},$$

which collapses the disk to the segment $(-1, 1)$ and has derivative $\partial f/\partial z(0) = 2/\pi i$.

We now turn to the analogue of the Koebe one-quarter theorem.

Theorem 2. *Each function $f \in S_H^0$ satisfies the inequality*

$$|f(z)| \geq \frac{1}{4} \frac{|z|}{(1+|z|)^2}, \qquad |z| < 1.$$

In particular, the range of f contains the disk $|w| < \frac{1}{16}$.

The proof will appeal to the method of extremal length, as developed in the Appendix. Recall first that the *module* of a ring domain Ω is $\mu(\Omega) = \frac{1}{2\pi} \log (r_2/r_1)$ if Ω can be mapped conformally onto an annulus $r_1 < |w| < r_2$. The relation to be invoked is

$$\frac{1}{\mu(\Omega)} = \inf_{\rho} \int\int_{\Omega} \rho(z)^2 \, dx \, dy,$$

where the infimum is taken over all admissible metrics ρ. An *admissible metric* is a measurable function $\rho(z) \geq 0$ defined on Ω such that

$$\int_{\gamma} \rho(z) \, |dz| \geq 1$$

for every rectifiable arc γ lying in Ω and joining the two boundary components. Thus, the module of Ω is the extremal length of this family of arcs. (See the Appendix for further information.)

Proof of Theorem. It is convenient to work with the dilation $f(z) = F(az)/a$ of a function $F \in S_H^0$ for some positive constant $a < 1$. Then $f \in S_H^0$ and f is a homeomorphism of the closed disk $\overline{\mathbb{D}}$. Let $\Omega = f(\mathbb{D})$ be the range of f and let Ω_ε be the ring domain obtained from Ω by deleting a small disk $\{w : |w| \leq \varepsilon\} \subset \Omega$. Let δ be the distance from the origin to the boundary curve $\partial\Omega$, and suppose without loss of generality that $-\delta \in \partial\Omega$. Let

$$\Delta = \mathbb{C}\backslash\{w : -\infty < w \leq -\delta\}$$

be the complex plane minus the ray from $-\delta$ to infinity and let

$$\Delta_\varepsilon = \Delta\backslash\{w : |w| \leq \varepsilon\}$$

be the Grötzsch ring bounded by this ray and the circle of radius ε. It is mapped to a slit disk, also called a Grötzsch ring (see Section 5.2), by an inversion. Note that $0 < \varepsilon < \delta$.

The first step is to establish the inequality $\mu(\Omega_\varepsilon) \leq \mu(\Delta_\varepsilon)$. This is not so easy, since it need not be true that $\Omega_\varepsilon \subset \Delta_\varepsilon$. One approach is to apply a circular symmetrization to Ω_ε to obtain a new domain $\Omega_\varepsilon^* \subset \Delta_\varepsilon$, then to invoke a general theorem of Pólya and Szegő [1] (see also Hayman [1], p. 75) to conclude that $\mu(\Omega_\varepsilon) \leq \mu(\Omega_\varepsilon^*) \leq \mu(\Delta_\varepsilon)$. However, it seems preferable to give a self-contained argument by extremal length (*cf.* Fuchs [1], p. 93) as follows.

Let Γ be the family of crosscuts (arcs joining the boundary components) in Δ_ε, and let C be the family of crosscuts in Ω_ε. Suppose now that ρ is an admissible metric for C. Extend ρ to Δ_ε by setting $\rho(z) = 0$ outside Ω_ε. Then $\rho_1(z) = \frac{1}{2}[\rho(z) + \rho(\overline{z})]$ is an admissible metric for Γ. To see this, let $\gamma \in \Gamma$ and consider

$$\int_\gamma \rho_1(z)\,|dz| = \frac{1}{2}\int_\gamma \rho(z)\,|dz| + \frac{1}{2}\int_{\overline{\gamma}} \rho(z)\,|dz| = \frac{1}{2}\int_{\gamma \cup \overline{\gamma}} \rho(z)\,|dz|.$$

Now $\gamma \cup \overline{\gamma}$ contains two arcs α and β in C, so

$$\int_\gamma \rho_1(z)\,|dz| \geq \frac{1}{2}\int_\alpha \rho(z)\,|dz| + \frac{1}{2}\int_\beta \rho(z)\,|dz| \geq 1.$$

Thus, ρ_1 is admissible for Γ, and

$$\frac{1}{\mu(\Delta_\varepsilon)} \leq \iint_{\Delta_\varepsilon} \rho_1(z)^2\,dx\,dy = \frac{1}{4}\iint_{\Delta_\varepsilon} \{\rho(z)^2 + 2\rho(z)\rho(\overline{z}) + \rho(\overline{z})^2\}\,dx\,dy$$

$$\leq \iint_{\Delta_\varepsilon} \rho(z)^2\,dx\,dy = \iint_{\Omega_\varepsilon} \rho(z)^2\,dx\,dy$$

by the Schwarz inequality, since Δ_ε is symmetric with respect to the real axis. Now take the infimum of the right-hand side over all admissible metrics ρ to conclude that $1/\mu(\Delta_\varepsilon) \leq 1/\mu(\Omega_\varepsilon)$, or $\mu(\Omega_\varepsilon) \leq \mu(\Delta_\varepsilon)$.

The next step is to estimate $\mu(\Omega_\varepsilon)$ from below. This can be done by judicious choice of an admissible metric. In making the calculation, it is convenient to identify Ω_ε with $f(\mathbb{D}_\varepsilon)$, where $\mathbb{D}_\varepsilon$ is the annulus defined by $\varepsilon < |z| < 1$. In view of the normalization $f(z) = z + O(|z|^2)$ of a function $f \in S_H^0$, the two regions are essentially the same for small ε. Rather than make this intuitive leap, however, we shall proceed as follows.

Fix a parameter $\beta > 1$ and observe that $|f(z)| = \beta\varepsilon + O(\varepsilon^2) > \varepsilon$ on the circle $|z| = \beta\varepsilon$ for sufficiently small $\varepsilon > 0$. Let ρ be an admissible metric for the family of crosscuts of $\mathbb{D}_{\beta\varepsilon}$, let $\rho(z) = 0$ outside $\mathbb{D}_{\beta\varepsilon}$, and define a metric $\tilde{\rho}$ on Ω_ε by

$$\tilde{\rho}(w) = \frac{\rho(z)}{|h'(z)| - |g'(z)|}, \qquad w = f(z),$$

where $f = h + \bar{g}$ is the canonical representation of f. Let $\tilde{\gamma}$ be a crosscut of Ω_ε and let γ be the restriction of its preimage $f^{-1}(\tilde{\gamma})$ to a crosscut of $\mathbb{D}_{\beta\varepsilon}$. Then

$$\int_{\tilde{\gamma}} \tilde{\rho}(w) \, |dw| = \int_\gamma \frac{\rho(z)|\frac{\partial f}{\partial s}(z)|}{|h'(z)| - |g'(z)|} \, |dz|,$$

where $\partial f / \partial s$ denotes the directional derivative of f along γ. But

$$\left| \frac{\partial f}{\partial s}(z) \right| \geq |h'(z)| - |g'(z)|,$$

so

$$\int_{\tilde{\gamma}} \tilde{\rho}(w) \, |dw| \geq \int_\gamma \rho(z) \, |dz| \geq 1,$$

and $\tilde{\rho}$ is admissible for the family C of crosscuts in Ω_ε. Therefore,

$$\frac{1}{\mu(\Omega_\varepsilon)} \leq \iint_{\Omega_\varepsilon} \tilde{\rho}(w)^2 \, du \, dv = \iint_{\mathbb{D}_{\beta\varepsilon}} \frac{\rho(z)^2 J_f(z)}{[|h'(z)| - |g'(z)|]^2} \, dx \, dy,$$

where $J_f(z) = |h'(z)|^2 - |g'(z)|^2$ is the Jacobian of f.

Recall now that f was assumed to have the form $f(z) = F(az)/a$ for some positive number $a < 1$. Consequently,

$$\frac{J_f(z)}{[|h'(z)| - |g'(z)|]^2} = \frac{|h'(z)| + |g'(z)|}{|h'(z)| - |g'(z)|} = \frac{1 + |\omega(z)|}{1 - |\omega(z)|} \leq \frac{1 + a|z|}{1 - a|z|}$$

by the Schwarz lemma, where $\omega = h'/g'$ is the dilatation of f. Thus, the inequality

$$\frac{1}{\mu(\Omega_\varepsilon)} \leq \iint_{\mathbb{D}_{\beta\varepsilon}} \frac{1 + a|z|}{1 - a|z|} \rho(z)^2 \, dx \, dy$$

holds for any metric ρ that is admissible for the family of crosscuts in $\mathbb{D}_{\beta\varepsilon}$.

Now choose the admissible metric

$$\rho(z) = \frac{1 - ar}{1 + ar} \frac{1}{r} \left\{ \int_{\beta\varepsilon}^1 \frac{1 - at}{1 + at} \frac{1}{t} \, dt \right\}^{-1}, \qquad r = |z|.$$

Through straightforward integration this leads to the inequality

$$\frac{1}{\mu(\Omega_\varepsilon)} \leq 2\pi \left\{ \int_{\beta\varepsilon}^1 \frac{1-at}{1+at} \frac{1}{t} \, dt \right\}^{-1},$$

or

$$\mu(\Omega_\varepsilon) \geq \frac{1}{2\pi} \left(2\log \frac{1+a\beta\varepsilon}{1+a} - \log \beta\varepsilon \right).$$

On the other hand, the function $\psi(z) = 4\delta k(z)$ maps $\mathbb{D}$ onto $\mathbb{C}\backslash\{-\infty < w \leq -\delta\}$, and so $\psi(\mathbb{D}_{\varepsilon/4\delta})$ agrees approximately with Δ_ε when ε is small. Since the module of a ring domain is conformally invariant, this gives the approximate formula

$$\mu(\Delta_\varepsilon) \approx \mu(\mathbb{D}_{\varepsilon/4\delta}) = \frac{1}{2\pi} \log \frac{4\delta}{\varepsilon}.$$

More rigorously, the module of the Grötzsch ring Δ_ε can be expressed precisely in terms of complete elliptic integrals (see Section 5.2), and asymptotic formulas give

$$\mu(\Delta_\varepsilon) = \frac{1}{2\pi} \log \frac{4\delta}{\varepsilon} + O(\varepsilon), \qquad \varepsilon \to 0.$$

Recalling that $\mu(\Omega_\varepsilon) \leq \mu(\Delta_\varepsilon)$, we conclude that

$$2 \log \frac{1+a\beta\varepsilon}{1+a} - \log \beta\varepsilon \leq \log \frac{4\delta}{\varepsilon} + O(\varepsilon),$$

which implies as ε tends to 0 that $\delta \geq \frac{1}{4\beta}(1+a)^{-2}$. Letting β tend to 1, we deduce finally that $\delta \geq \frac{1}{4}(1+a)^{-2}$. However, $|f(z)| \geq \delta$ on the unit circle, by the definition of δ. Thus, because $f(z) = F(az)/a$, it follows that

$$|F(az)| \geq \frac{1}{4}a(1+a)^{-2} \qquad \text{for } |z| = 1.$$

Since the function F in S_H^0 and the number a $(0 < a < 1)$ were chosen arbitrarily, the theorem is proved. ∎

6.3. Estimation of $|a_2|$

For analytic univalent functions $f(z) = z + a_2 z^2 + \cdots$ of class S, an old problem of key importance was to find the sharp bound for the second coefficient a_2. In 1916, by elementary devices that are now quite familiar, Bieberbach proved $|a_2| \leq 2$ and applied the result to obtain the sharp forms of Koebe's growth and distortion theorems, and of Koebe's covering theorem (the Koebe one-quarter theorem). (An exposition may be found, for instance, in Duren [2], Ch. 2.) The corresponding problem for harmonic mappings, to find the sharp bound for $|a_2|$ in the class S_H^0, is still unresolved. Its solution would have similar geometric applications, as described in Section 6.4.

Recall that each function f in S_H^0 has a canonical representation $f = h + \bar{g}$, where $h(z) = z + a_2 z^2 + \cdots$ and $g(z) = b_2 z^2 + \cdots$. As mentioned in Section 5.4, the harmonic Koebe function suggests the conjecture that $|a_2| \leq \frac{5}{2}$ for all $f \in S_H^0$. Although the best bound is unknown, Sheil-Small [3] was able to show that $|a_2| < 57$, improving on an earlier estimate by Clunie and Sheil-Small [1]. Here we shall present Sheil-Small's argument with small modifications leading to a slightly better bound.

Theorem. *The inequality $|a_2| < 49$ holds for every function in the class S_H^0.*

Proof. The point of departure is the analogue of Koebe's covering theorem for S_H^0, developed in the preceding section. It says that the range of each harmonic mapping $f \in S_H^0$ contains the disk $|w| < \frac{1}{16}$. Let Δ be the preimage of this disk under f, and observe that Δ contains the origin. Let φ be the (analytic) conformal mapping of $\mathbb{D}$ onto Δ with $\varphi(0) = 0$ and $\varphi'(0) > 0$. Then the composition $F = 16 f \circ \varphi$ is a harmonic mapping of $\mathbb{D}$ onto itself.

Now let $\varphi(z) = c_1 z + c_2 z^2 + \cdots$ and expand

$$f(\varphi(z)) = \varphi(z) + a_2 \, \varphi(z)^2 + \cdots + \overline{b_2} \, \overline{\varphi(z)}^2 + \cdots$$
$$= c_1 z + (c_2 + a_2 c_1^2) z^2 + \cdots + \overline{b_2} \overline{c_1}^2 \bar{z}^2 + \cdots.$$

Thus, with the notation

$$F(z) = 16 f(\varphi(z)) = \sum_{n=1}^{\infty} A_n z^n + \overline{\sum_{n=2}^{\infty} B_n z^n},$$

we have the formulas

$$A_1 = 16 c_1, \qquad A_2 = 16 \left(c_2 + a_2 c_1^2 \right).$$

Since F maps the unit disk harmonically onto itself, it follows that $|A_n| \leq \frac{1}{n}$, as shown in Section 4.5. In particular, $|A_2| \leq \frac{1}{2}$ or $|c_2 + a_2 c_1^2| \leq \frac{1}{32}$. Thus, we arrive at the estimate

$$|a_2| \leq |c_1|^{-2} \left\{ |c_2| + \frac{1}{32} \right\}. \qquad \blacksquare$$

On the other hand, a classical theorem of Pick (see Duren [2], p. 74) says that $|a_2| \leq 2 \left(1 - \frac{1}{M} \right)$ for bounded functions f in the class S with $|f(z)| < M$. When applied to the function φ, Pick's inequality gives

$$|c_2| \leq 2 \, |c_1| \, (1 - |c_1|).$$

Therefore, the estimate for $|a_2|$ (for functions in S_H^0) reduces to

$$|a_2| \leq \frac{1}{32} |c_1|^{-2} + 2 |c_1|^{-1} - 2.$$

Finally, the sharp form of the Heinz inequality (Section 4.4) gives the lower bound $|A_1|^2 \geq \frac{27}{4}\pi^{-2}$, or $|c_1|^2 \geq \frac{27}{1024}\pi^{-2}$, which leads to the estimate

$$|a_2| \leq \frac{32\pi}{27}(\pi + 6\sqrt{3}) - 2 < 48.4$$

Even if the radius $\frac{1}{16}$ in the covering theorem for S_H^0 were replaced by the conjectured sharp value $\frac{1}{6}$, the preceding proof would give only $|a_2| < 17$.

6.4. Growth and Distortion

A well-known result from the classical theory of analytic univalent functions is Bieberbach's coefficient bound $|a_2| \leq 2$ for functions $f(z) = z + a_2 z^2 + \cdots$ in the class S. This seemingly technical estimate has important geometric implications. Through various elementary transformations, it leads to sharp growth and distortion theorems, and it provides an easy proof of the Koebe one-quarter theorem (see, for instance, Duren [2], Ch. 2).

The situation for harmonic mappings is quite similar. For functions of class S_H it is conjectured that $|a_2| < 3$. Although the conjecture remains unsettled, we will now show that its truth would yield sharp growth estimates for functions in S_H^0 and would give the sharp form of the covering theorem, namely, that each such function includes the disk $|w| < \frac{1}{6}$ in its range. (A weaker version of the covering theorem, with radius $\frac{1}{16}$ instead of $\frac{1}{6}$, was established in Section 6.2.) The following theorem is due to Sheil-Small [3].

Theorem. *Let α be the supremum of $|a_2|$ among all functions $f \in S_H$. Then every function $f \in S_H^0$ satisfies the inequalities*

$$\frac{1}{2\alpha}\left[1 - \left(\frac{1-r}{1+r}\right)^\alpha\right] \leq |f(z)| \leq \frac{1}{2\alpha}\left[\left(\frac{1+r}{1-r}\right)^\alpha - 1\right], \qquad r = |z| < 1.$$

In particular, the range of each function $f \in S_H^0$ contains the disk $|w| < \frac{1}{2\alpha}$.

Before turning to the proof, let us observe that if $\alpha = 3$ as conjectured, then both estimates are best possible. Recall (from Section 5.3) that the harmonic Koebe function belongs to S_H^0 and has the form $K = H + \overline{G}$, where

$$H(z) = \frac{z - \frac{1}{2}z^2 + \frac{1}{6}z^3}{(1-z)^3}, \qquad G(z) = \frac{\frac{1}{2}z^2 + \frac{1}{6}z^3}{(1-z)^3}.$$

It is therefore seen that

$$K(r) = \frac{1}{6}\left[\left(\frac{1+r}{1-r}\right)^3 - 1\right] \quad \text{and} \quad K(-r) = \frac{1}{6}\left[\left(\frac{1-r}{1+r}\right)^3 - 1\right]$$

for $0 \le r < 1$.

Proof of Theorem. Choosing $f = h + \overline{g} \in S_H$ and fixing $\zeta \in \mathbb{D}$, we apply a disk automorphism to obtain the function

$$F(z) = \frac{f\left(\frac{z+\zeta}{1+\overline{\zeta}z}\right) - f(\zeta)}{(1 - |\zeta|^2)h'(\zeta)} = H(z) + \overline{G(z)},$$

which again belongs to S_H. With the notation

$$H(z) = z + A_2(\zeta)z^2 + A_3(\zeta)z^3 + \cdots,$$

a simple calculation gives

$$A_2(\zeta) = \frac{1}{2}\left\{(1 - |\zeta|^2)\frac{h''(\zeta)}{h'(\zeta)} - 2\overline{\zeta}\right\}.$$

But $|A_2(\zeta)| \le \alpha$ by definition of α, which implies that

$$\frac{2r^2 - 2\alpha r}{1 - r^2} \le \text{Re}\left\{\frac{zh''(z)}{h'(z)}\right\} \le \frac{2r^2 + 2\alpha r}{1 - r^2}, \qquad |z| = r < 1.$$

This inequality can be recast in the form

$$\frac{2r - 2\alpha}{1 - r^2} \le \frac{\partial}{\partial r}\{\log |h'(re^{i\theta})|\} \le \frac{2r + 2\alpha}{1 - r^2}.$$

After integration from 0 to r and exponentiation, we arrive at the estimate

$$\frac{(1 - r)^{\alpha-1}}{(1 + r)^{\alpha+1}} \le |h'(z)| \le \frac{(1 + r)^{\alpha-1}}{(1 - r)^{\alpha+1}}, \qquad |z| = r < 1.$$

Suppose now that $f = h + \overline{g} \in S_H^0$, so that $f + c\overline{f} \in S_H$ for every choice of complex constant $c \in \mathbb{D}$. Then, by what we have just proved,

$$\frac{(1-r)^{\alpha-1}}{(1+r)^{\alpha+1}} \leq |h'(z) + cg'(z)| \leq \frac{(1+r)^{\alpha-1}}{(1-r)^{\alpha+1}}.$$

This shows in particular that

$$\frac{(1-r)^{\alpha-1}}{(1+r)^{\alpha+1}} \leq |h'(z)| - |g'(z)|$$

and

$$|h'(z)| + |g'(z)| \leq \frac{(1+r)^{\alpha-1}}{(1-r)^{\alpha+1}}.$$

The last inequality shows that

$$|f(z)| = \left| \int_\Gamma \frac{\partial f}{\partial \zeta} d\zeta + \frac{\partial f}{\partial \overline{\zeta}} d\overline{\zeta} \right| \leq \int_\Gamma (|h'(\zeta)| + |g'(\zeta)|) |d\zeta|$$

$$\leq \int_0^r \frac{(1+\rho)^{\alpha-1}}{(1-\rho)^{\alpha+1}} d\rho = \frac{1}{2\alpha} \left[\left(\frac{1+r}{1-r} \right)^\alpha - 1 \right]$$

for $|z| = r < 1$, where Γ is the radial line segment from 0 to z. Next let Γ be the preimage under f of the radial segment from 0 to $f(z)$. Then

$$|f(z)| = \int_\Gamma \left| \frac{\partial f}{\partial \zeta} d\zeta + \frac{\partial f}{\partial \overline{\zeta}} d\overline{\zeta} \right| \geq \int_\Gamma (|h'(\zeta)| - |g'(\zeta)|) |d\zeta|$$

$$\geq \int_0^r \frac{(1-\rho)^{\alpha-1}}{(1+\rho)^{\alpha+1}} d\rho = \frac{1}{2\alpha} \left[1 - \left(\frac{1-r}{1+r} \right)^\alpha \right],$$

which completes the proof. ∎

The theorem is actually a special case of a more general result, as Sheil-Small [3] emphasizes. The only properties of the class S_H essential to the proof are its affine and linear invariance. Thus, the theorem remains valid for any subclass of S_H that is invariant under normalized affine transformations

$$f \to \frac{f + c\overline{f}}{1 + cb_1}, \qquad c \in \mathbb{D},$$

and all disk automorphisms $f \longmapsto F$ as performed above. For instance, the family C_H of convex mappings has the required invariance properties, so the theorem applies to this subclass. The second coefficients of functions in

C_H satisfy the sharp inequality $|a_2| < 2$ (*cf.* Section 3.6), so $\alpha = 2$ and the theorem (suitably generalized) gives the bounds

$$\frac{r}{(1+r)^2} \le |f(z)| \le \frac{r}{(1-r)^2}, \qquad |z| = r < 1,$$

for every function $f \in C_H^0$. An alternate derivation of the upper bound uses the coefficient estimates (Section 3.6, Theorem 2) for functions in C_H^0. However, the lower bound is clearly not sharp, since we saw in Section 3.6 that the disk $|w| < \frac{1}{2}$ is contained in the range of each function of class C_H^0. It seems likely that the sharp bounds are attained by the standard half-plane mapping

$$L(z) = \text{Re}\,\{\ell(z)\} + i\,\text{Im}\,\{k(z)\},$$

as developed in Section 3.4. This appears to be an open problem. In any event, it can be shown that the upper bound is correct in order of magnitude, since the function L has a maximum modulus $M_\infty(r, L)$ that grows like $(1 - r)^{-2}$. Indeed, if we write

$$L(re^{i\theta}) = \frac{r\cos\theta - r^2}{1 - 2r\cos\theta + r^2} + i\,\frac{r(1 - r^2)\sin\theta}{(1 - 2r\cos\theta + r^2)^2}$$
$$= \phi_r(\theta) + i\psi_r(\theta),$$

we can see that $\phi_r(1 - r) \sim \frac{1}{2}(1 - r)^{-1}$ and $\psi_r(1 - r) \sim \frac{1}{2}(1 - r)^{-2}$ as $r \to 1$. Thus, $|L(re^{i(1-r)})| \sim \frac{1}{2}(1 - r)^{-2}$, and it follows that

$$\limsup_{r\to 1} (1 - r)^2 M_\infty(r, L) \ge \frac{1}{2}.$$

The generalized theorem does not apply directly to starlike mappings, since the subclass of S_H consisting of functions starlike with respect to the origin is not linearly invariant. The center of starlikeness shifts under disk automorphism. However, the larger subclass of close-to-convex mappings is linearly invariant, and the sharp inequality $|a_2| \le 3$ for this family was established by Clunie and Sheil-Small [1]. The theorem therefore gives the bounds

$$\frac{1}{3}\frac{3r + r^3}{(1+r)^3} \le |f(z)| \le \frac{1}{3}\frac{3r + r^3}{(1-r)^3}, \qquad |z| = r < 1,$$

for all close-to-convex functions, hence for all starlike functions in S_H^0. As noted earlier, the harmonic Koebe function $K(z)$, a starlike function in S_H^0, shows that the bounds are sharp. In Section 6.7 we shall verify the upper bound directly as a consequence of the sharp coefficient inequalities for starlike functions in S_H^0.

6.5. Marty Relation

Variational methods offer a powerful approach to extremal problems in families of analytic univalent functions. It would be desirable to adapt such methods to harmonic mappings, but serious obstacles arise because harmonic mappings are not preserved under composition, and because a harmonic mapping is not determined by its range. Nevertheless, a harmonic function of an analytic function is always harmonic. For this reason an elementary method known as the *Marty variation* extends at once to harmonic mappings. The method was introduced by Felix Marty in 1934 to study the coefficient problem for analytic univalent functions (see Duren [2], Sec. 2.9). Marty found that if $f(z) = z + a_2 z^2 + \cdots$ is any function of class S that maximizes the real part of the nth coefficient, then

$$(n + 1)a_{n+1} - 2a_2 a_n - (n - 1)\,\overline{a_{n-1}} = 0.$$

This is known as the *Marty relation*.

A slight modification of the Marty variation leads to analogues of the Marty relation for harmonic mappings. Let $f = h + \overline{g}$ belong to S_H^0, where $h(z) = z + a_2 z^2 + \cdots$ and $g(z) = b_2 z^2 + \cdots$. Let $\zeta \in \mathbb{D}$ be a small complex parameter and consider the perturbation

$$F(z) = \frac{f\left(\frac{z+\zeta}{1+\overline{\zeta}z}\right) - f(\zeta)}{(1 - |\zeta|^2)h'(\zeta)} = H(z) + \overline{G(z)},$$

where $H(z) = z + A_2(\zeta)z^2 + \cdots$ and $G(z) = B_1(\zeta)z + B_2(\zeta)z^2 + \cdots$. Note that $F \in S_H$ by construction. Calculations give

$$\frac{z+\zeta}{1+\overline{\zeta}z} = z + (\zeta - \overline{\zeta}z^2) + O(|\zeta|^2)$$

and more generally

$$\left(\frac{z+\zeta}{1+\overline{\zeta}z}\right)^k = z^k + k(\zeta - \overline{\zeta}z^2)z^{k-1} + O(|\zeta|^2), \qquad k = 1, 2, \cdots.$$

Thus,

$$h\left(\frac{z+\zeta}{1+\overline{\zeta}z}\right) = \sum_{k=1}^{\infty} a_k[z^k + k(\zeta - \overline{\zeta}z^2)z^{k-1}] + O(|\zeta|^2),$$

where $a_1 = 1$, and

$$g\left(\frac{z+\zeta}{1+\overline{\zeta}z}\right) = \sum_{k=2}^{\infty} b_k[z^k + k(\zeta - \overline{\zeta}z^2)z^{k-1}] + O(|\zeta|^2),$$

Combining these expansions with

$$(1 - |\zeta|^2)h'(\zeta) = 1 + 2a_2\zeta + O(|\zeta|^2),$$

one arrives at the asymptotic formulas

$$A_n(\zeta) = a_n + [(n + 1)a_{n+1} - 2a_2 a_n]\zeta - (n - 1)a_{n-1}\overline{\zeta} + O(|\zeta|^2)$$

for $n = 2, 3, \ldots$, where $a_1 = 1$; and

$$B_n(\zeta) = b_n + (n + 1)b_{n+1}\zeta - [2\overline{a_2}b_n + (n - 1)b_{n-1}]\overline{\zeta} + O(|\zeta|^2)$$

for $n = 1, 2, \ldots$, where $b_0 = b_1 = 0$.

The function $F \in S_H$ is now composed with the standard affine mapping to produce the desired variation $f^* \in S_H^0$. Specifically,

$$f^*(z) = \frac{F(z) - \overline{B_1(\zeta)}\,\overline{F(z)}}{1 - |B_1(\zeta)|^2} = h^*(z) + \overline{g^*(z)},$$

where $h^*(z) = z + a_2^* z^2 + \cdots$ and $g^*(z) = b_2^* z^2 + \cdots$. Further calculations lead to the expressions

$$a_n^* = a_n + [(n + 1)a_{n+1} - 2a_2 a_n]\zeta - [2\overline{b_2}b_n + (n - 1)a_{n-1}]\overline{\zeta} + O(|\zeta|^2)$$

and

$$b_n^* = b_n + [(n + 1)b_{n+1} - 2b_2 a_n]\zeta - [2\overline{a_2}b_n + (n - 1)b_{n-1}]\overline{\zeta} + O(|\zeta|^2)$$

for $n = 2, 3, \ldots$.

The variation just constructed will now be applied to study the coefficient conjectures for the class S_H^0, namely, that the harmonic Koebe function maximizes $|a_n|$ and $|b_n|$ for each $n \geq 2$ (see Section 5.4). For fixed $n \geq 2$, suppose first that a function $f \in S_H^0$ has a coefficient a_n of largest modulus. After a rotation it may be assumed that $a_n > 0$, so that f has a coefficient a_n of largest real part. Since $f^* \in S_H^0$, it follows from the extremal property of f that $\text{Re}\,\{a_n^*\} \leq \text{Re}\,\{a_n\}$, or

$$\text{Re}\,\left\{[(n + 1)a_{n+1} - 2a_2 a_n - 2b_2\overline{b_n} - (n - 1)\,\overline{a_{n-1}}]\zeta + O(|\zeta|^2)\right\} \leq 0.$$

Dividing by ζ and letting ζ tend to 0 along an arbitrary ray, we conclude that

$$(n + 1)a_{n+1} - 2a_2 a_n - 2b_2\overline{b_n} - (n - 1)\,\overline{a_{n-1}} = 0.$$

This is the analogue of the Marty relation. Similarly, if f has a coefficient b_n of maximum real part, the same reasoning leads to the necessary condition

$$(n + 1)b_{n+1} - 2a_2\overline{b_n} - 2b_2 a_n - (n - 1)\overline{b_{n-1}} = 0.$$

In each case it may be checked that the harmonic Koebe function, with coefficients $a_k = \frac{1}{6}(2k + 1)(k + 1)$ and $b_k = \frac{1}{6}(2k - 1)(k - 1)$, actually satisfies the corresponding Marty relation and therefore still qualifies as a possible extremal function. In other words, the Marty variation gives evidence in favor of the conjectured inequalities $|a_n| \leq \frac{1}{6}(2n + 1)(n + 1)$ and $|b_n| \leq \frac{1}{6}(2n - 1)(n - 1)$ for all functions in the class S_H^0.

On the other hand, it may be observed that for each $n \geq 2$ the function $f(z) = z + \frac{1}{n}\bar{z}^n$ belongs to S_H^0 and satisfies the Marty relation for the extremal problem of maximizing $\text{Re}\,\{b_n\}$ (all four terms vanish), yet for no $n \geq 3$ is its coefficient $b_n = \frac{1}{n}$ the largest possible. Indeed, $\frac{1}{n} < \frac{1}{6}(2n - 1)(n - 1)$ for each $n \geq 3$.

6.6. Typically Real Functions

Although the coefficient conjectures

$$|a_n| \leq \frac{1}{6}(2n + 1)(n + 1) \qquad \text{and} \qquad |b_n| \leq \frac{1}{6}(2n - 1)(n - 1)$$

are still open for the full class S_H^0 of normalized harmonic mappings, the bounds can be established for certain subclasses. In this section and the next, we shall verify the conjectured bounds for typically real functions and for starlike mappings, respectively.

Recall first that a function analytic in $\mathbb{D}$ is said to be typically real if its values are real on the real axis and nonreal elsewhere. Every typically real analytic function has real coefficients. Let T denote the class of analytic typically real functions φ for which $\varphi(0) = 0$ and $\varphi'(0) = 1$. The univalent functions $\varphi \in S$ with real coefficients form a proper subclass of T. It is not difficult to prove (see Duren [2], p. 58) that the coefficients of every function

$$\varphi(z) = z + c_2 z^2 + c_3 z^3 + \cdots$$

of class T satisfy the inequality $|c_n| \leq n$. The Koebe function belongs to T and shows that the bound is sharp for each n.

The harmonic typically real functions are defined in a similar way. A complex-valued function f harmonic in $\mathbb{D}$ is said to be *typically real* provided $f(z)$ is real if and only if z is real. The class T_H consists of all sense-preserving typically real harmonic functions $f = h + \bar{g}$ with $h(0) = g(0) = 0$, $|h'(0)| = 1$, and $f(r) > 0$ for $0 < r < 1$. Here it is not required that $h'(0) = 1$. The subclass of T_H with $g'(0) = 0$ is denoted by T_H^0. A typically real harmonic function need not be univalent. However, every function $f \in S_H$ with real coefficients is typically real and belongs to the class T_H. Indeed, if $f = h + \overline{g}$ has all of its coefficients a_n and b_n real, then

$$\overline{f(z)} = \overline{h(z)} + g(z) = f(\overline{z}) \qquad \text{for all} \quad z \in \mathbb{D},$$

while $f(z)$ is real if and only if $\overline{f(z)} = f(z)$. Thus, $f(z) = f(\overline{z})$ wherever $f(z)$ is real. But if f is univalent, this can happen only for $z = \overline{z}$, which means that $f(z)$ is real only where z is real. Thus, $f \in T_H$.

If $f = h + \overline{g} \in T_H$, then $\text{Im}\,\{f(z)\} > 0$ for $\text{Im}\,\{z\} > 0$, while $\text{Im}\,\{f(z)\} < 0$ for $\text{Im}\,\{z\} < 0$. Since

$$\text{Im}\,\{f(z)\} = \text{Im}\,\{h(z) + \overline{g(z)}\} = \text{Im}\,\{h(z) - g(z)\},$$

it is clear that f is typically real if and only if the analytic function $\varphi = h - g$ is typically real. Since an analytic typically real function has real coefficients, it follows that $a_n - b_n$ is real $(n = 1, 2, \ldots)$ for every function $f \in T_H$. Thus, $\overline{a_1} - \overline{b_1}$ is real, and $|b_1| < |a_1| = 1$, so the function

$$f_0(z) = \frac{\overline{a_1} f(z) - \overline{b_1}\,\overline{f(z)}}{1 - |b_1|^2} = h_0(z) + \overline{g_0(z)}$$

is a sense-preserving harmonic function that is seen to be typically real. Furthermore, the construction ensures that $f_0(0) = 0$, $h_0'(0) = 1$, and $g_0'(0) = 0$, so $f_0 \in T_H^0$.

This last formula can be inverted to give

$$f(z) = a_1 f_0(z) + \overline{b_1}\,\overline{f_0(z)}.$$

Moreover, if any function f has this form for some function $f_0 \in T_H^0$, and for some constants a_1 and b_1 with $|b_1| < |a_1| = 1$ and $a_1 + \overline{b_1} > 0$, then it is clear that $f \in T_H$.

The following theorems are due to Clunie and Sheil-Small [1].

Theorem 1. *If $f = h + \overline{g} \in T_H^0$, then $a_1 = 1$,*

$$|a_n| \le \frac{1}{6}(2n + 1)(n + 1), \qquad |b_n| \le \frac{1}{6}(2n - 1)(n - 1),$$

and

$$||a_n| - |b_n|| \le n \qquad \text{for} \quad n = 2, 3, \ldots.$$

Theorem 2. *If $f \in T_H$, then*

$$|a_n| < \frac{1}{3}(2n^2 + 1) \qquad \text{and} \qquad |b_n| < \frac{1}{3}(2n^2 + 1)$$

for $n = 2, 3, \ldots$. Each bound is sharp.

Deduction of Theorem 2. Before turning to the proof of Theorem 1, we observe that Theorem 2 is an immediate consequence. Indeed, as we have just remarked, each function $f \in T_H$ has the form $f = a_1 f_0 + \overline{b_1} \overline{f_0}$ for some function $f_0 \in T_H^0$ and for complex constants a_1 and b_1 with $|b_1| < |a_1| = 1$ and $a_1 + \overline{b_1} > 0$. Thus, by Theorem 1 the coefficients of f satisfy the inequalities

$$|a_n| \le |a_1| \frac{1}{6}(2n + 1)(n + 1) + |b_1| \frac{1}{6}(2n - 1)(n - 1)$$

$$< \frac{1}{6}(2n + 1)(n + 1) + \frac{1}{6}(2n - 1)(n - 1) = \frac{1}{3}(2n^2 + 1)$$

and

$$|b_n| \le |a_1| \frac{1}{6}(2n - 1)(n - 1) + |b_1| \frac{1}{6}(2n + 1)(n + 1) < \frac{1}{3}(2n^2 + 1).$$

To see that the bounds are sharp (but not attained by any function $f \in T_H$), choose f_0 to be the harmonic Koebe function, choose $a_1 = 1$ and $0 < b_1 < 1$, and let $b_1 \to 1$. The resulting limit function is

$$2 \operatorname{Re} \left\{ \frac{z + \frac{1}{3}z^3}{(1 - z)^3} \right\}.$$

Proof of Theorem 1. Since $b_1 = 0$, the requirement that $f(r) > 0$ for $0 < r < 1$ shows that $a_1 > 0$, so $a_1 = 1$. Consequently, the analytic function $\varphi = h - g$ belongs to the class T, so

$$\big||a_n| - |b_n|\big| \le |a_n - b_n| \le n$$

by the known coefficient bound for analytic typically real functions.

Next observe that the dilatation $\omega = g'/h'$ satisfies $|\omega(z)| \le |z|$, by the Schwarz lemma, since f is sense-preserving and $g'(0) = 0$. Writing

$$g'(z) = \omega(z)h'(z) = \omega(z)(\varphi'(z) + g'(z)),$$

we see that

$$g'(z) = \frac{\omega(z)}{1 - \omega(z)} \varphi'(z).$$

But the function $\omega/(1 - \omega)$ is subordinate to the convex function $z/(1 - z)$, so an appeal to Lemma 3 of Section 3.6 shows that the coefficients of $g'(z)$ are dominated in modulus by those of the function

$$\frac{z}{1 - z} \frac{1 + z}{(1 - z)^3} = z + \frac{1}{6} \sum_{n=2}^{\infty} n(n + 1)(2n + 1)z^n.$$

More specifically, $|2b_2| \leq 1$ and

$$|(n+1)b_{n+1}| \leq \frac{1}{6}n(n+1)(2n+1), \qquad n = 2, 3, \ldots,$$

or

$$|b_n| \leq \frac{1}{6}(2n-1)(n-1), \qquad n = 2, 3, \ldots.$$

The final estimate is now obtained by combining the two previous results:

$$|a_n| \leq ||a_n| - |b_n|| + |b_n| \leq n + \frac{1}{6}(2n-1)(n-1) = \frac{1}{6}(2n+1)(n+1).$$

This concludes the proof.   ■

It may be remarked that the proof actually has little to do with the assumption that f is typically real. The essential ingredient is the bound $|c_n| \leq n$ for the coefficients of the function $\varphi = h - g$. Thus, the same proof will apply whenever $f = h + \overline{g} \in S_H^0$ and $\varphi = h - g$ is *univalent*. This will be the case, in particular, when f is convex in one direction, in view of the shearing theorem (Theorem 1 of Section 3.4). We have therefore proved the following additional theorem.

Theorem 3. *If $f = h + \overline{g} \in S_H^0$ and its range is convex in one direction, then*

$$|a_n| \leq \frac{1}{6}(2n+1)(n+1), \qquad |b_n| \leq \frac{1}{6}(2n-1)(n-1),$$

and

$$||a_n| - |b_n|| \leq n \qquad \text{for } n = 2, 3, \ldots.$$

6.7. Starlike Functions

A sense-preserving harmonic mapping $f \in S_H$ is said to be *starlike* if its range is starlike with respect to the origin. This means that the whole range can be "seen" from the origin. In other words, if some point $w_0 = f(z_0)$ is in the range of f, then so is the entire radial segment from 0 to w_0. If f has a smooth extension to the closed disk, an equivalent requirement is that $\arg\{f(e^{i\theta})\}$ be a nondecreasing function of θ, or that

$$\frac{d}{d\theta} \arg\{f(e^{i\theta})\} \geq 0.$$

For analytic functions f, this condition takes the familiar form

$$\text{Re}\left\{\frac{zf'(z)}{f(z)}\right\} > 0, \qquad z \in \mathbb{D}.$$

A classical result known as Alexander's theorem (1915) asserts that f is convex if and only if the function $zf'(z)$ is starlike.

Continuing the theme of the last section, our aim is now to establish the sharp coefficient bounds for starlike harmonic functions of class S_H^0. Along the way, we shall see that Alexander's theorem has a partial extension to harmonic mappings. The following theorem is due to Sheil-Small [3].

Theorem. *The coefficients of every starlike function $f \in S_H^0$ satisfy the sharp inequalities*

$$|a_n| \le \frac{1}{6}(2n + 1)(n + 1), \qquad |b_n| \le \frac{1}{6}(2n - 1)(n - 1),$$

and

$$||a_n| - |b_n|| \le n \qquad \text{for } n = 2, 3, \ldots.$$

Since starlikeness is an affine-invariant property, the argument used in the previous section for typically real functions now allows us to deduce the sharp bounds for coefficients of starlike functions of class S_H. These estimates are actually implied by a more general result of Clunie and Sheil-Small [1], who found the corresponding bounds for close-to-convex functions of class S_H.

Corollary 1. *The coefficients of starlike functions of class S_H satisfy the inequalities*

$$|a_n| < \frac{1}{3}(2n^2 + 1) \qquad \text{and} \qquad |b_n| < \frac{1}{3}(2n^2 + 1)$$

for $n = 2, 3, \ldots$. Each bound is sharp, but none is attained.

Another consequence of the theorem is the sharp upper bound on the growth of a starlike harmonic mapping, which was already stated in Section 6.4.

Corollary 2. *The sharp inequality*

$$|f(z)| \le \frac{1}{3}\frac{3r + r^3}{(1 - r)^3}, \qquad |z| = r < 1$$

holds for every starlike function $f \in S_H^0$. Equality occurs for the harmonic Koebe function.

Proof. The theorem gives

$$|f(z)| \le \sum_{n=1}^{\infty} |a_n| r^n + \sum_{n=1}^{\infty} |b_n| r^n$$

$$\le \frac{1}{6} \sum_{n=1}^{\infty} (2n+1)(n+1) r^n + \frac{1}{6} \sum_{n=1}^{\infty} (2n-1)(n-1) r^n$$

$$= \frac{1}{3} \sum_{n=1}^{\infty} (2n^2+1) r^n = \frac{1}{3} \frac{3r+r^3}{(1-r)^3}. \qquad \blacksquare$$

The proof of the theorem will appeal to the following lemma, which represents a partial extension of Alexander's theorem to harmonic mappings.

Lemma. *If* $f = h + \overline{g} \in S_H$ *is a starlike function, and if H and G are the analytic functions defined by*

$$zH'(z) = h(z), \qquad zG'(z) = -g(z), \qquad H(0) = G(0) = 0,$$

then $F = H + \overline{G}$ *is a convex function of class* C_H.

Proof of Lemma. An application of the approximation theorem (Section 2.6) shows no generality is lost in assuming that f has a smooth extension to $\overline{\mathbb{D}}$ and that the boundary function gives a one-to-one sense-preserving mapping of the unit circle onto a curve starlike with respect to the origin. A simple calculation gives $\frac{d}{d\theta} F(e^{i\theta}) = if(e^{i\theta})$, so that

$$\frac{d}{d\theta} \arg \left\{ \frac{d}{d\theta} F(e^{i\theta}) \right\} = \frac{d}{d\theta} \arg \{ f(e^{i\theta}) \} \ge 0,$$

by the starlikeness of f. But this shows that F maps the unit circle onto a convex curve. It then follows from the Radó–Kneser–Choquet theorem that F maps $\mathbb{D}$ univalently onto a convex region. Hence, $F \in C_H$. $\qquad \blacksquare$

Proof of Theorem. By the lemma, the associated harmonic function $F = H + \overline{G}$ is of class C_H^0. Appealing now to Lemma 2 of Section 3.6, we conclude that

$$\mathrm{Re} \, \{ (e^{i\alpha} H'(z) + e^{-i\alpha} G'(z))(e^{i\beta} - e^{-i\beta} z^2) \} > 0$$

for some choice of angles α and β, and for all $z \in \mathbb{D}$. Equivalently,

$$\mathrm{Re} \left\{ \left(e^{i\alpha} \frac{h(z)}{z} - e^{-i\alpha} \frac{g(z)}{z} \right) (e^{i\beta} - e^{-i\beta} z^2) \right\} > 0$$

or

$$\text{Re} \left\{ \frac{e^{i\beta} - e^{-i\beta} z^2}{z} \sum_{n=1}^{\infty} \left(e^{i\alpha} a_n - e^{-i\alpha} b_n \right) z^n \right\} > 0,$$

where $a_1 = 1$ and $b_1 = 0$, since $f \in S_H^0$. This last result says that

$$\sum_{n=1}^{\infty} \left(e^{i\alpha} a_n - e^{-i\alpha} b_n \right) z^n = \frac{z}{e^{i\beta} - e^{-i\beta} z^2} P(z),$$

where P is an analytic function with the properties $|P(0)| = 1$ and $\text{Re}\{P(z)\} > 0$ in $\mathbb{D}$. But the first factor is a rotation of the function $z/(1 - z^2)$, so it follows from the Herglotz theorem (*cf.* Section 3.6, Lemma 1) that the coefficients of the power series are dominated in modulus by those of the function

$$\frac{z}{1 - z^2} \frac{1 + z}{1 - z} = \frac{z}{(1 - z)^2} = \sum_{n=1}^{\infty} n z^n.$$

In other words,

$$\left| e^{i\alpha} a_n - e^{-i\alpha} b_n \right| \le n, \qquad n = 1, 2, \ldots.$$

Thus, $||a_n| - |b_n|| \le n$.

For the other estimates, note first that $g'(z) = \omega(z) h'(z)$, where $|\omega(z)| \le |z|$. But differentiation gives

$$e^{i\alpha} h'(z) - e^{-i\alpha} g'(z) = \frac{d}{dz} \left\{ \frac{z}{e^{i\beta} - e^{-i\beta} z^2} P(z) \right\},$$

so that

$$g'(z) = \frac{\omega(z)}{e^{i\alpha} - e^{-i\alpha} \omega(z)} \frac{d}{dz} \left\{ \frac{z}{e^{i\beta} - e^{-i\beta} z^2} P(z) \right\}.$$

In view of the subordination principle (Section 3.6, Lemma 3), this shows that the coefficients of g' are dominated by those of the function

$$\frac{z}{1 - z} \frac{d}{dz} \left\{ \frac{z}{1 - z^2} \frac{1 + z}{1 - z} \right\} = \frac{z(1 + z)}{(1 - z)^4} = \frac{1}{6} \sum_{n=1}^{\infty} n(n + 1)(2n + 1) z^n.$$

In other words, $|n b_n| \le \frac{1}{6}(n - 1)n(2n - 1)$, or $|b_n| \le \frac{1}{6}(2n - 1)(n - 1)$ for $n = 2, 3, \ldots$. It now follows that

$$|a_n| \le |b_n| + ||a_n| - |b_n|| \le \frac{1}{6}(2n - 1)(n - 1) + n = \frac{1}{6}(2n + 1)(n + 1).$$

The harmonic Koebe function, which is clearly starlike, shows that all inequalities in the theorem are sharp.   ∎

Finally, we remark that Alexander's theorem does not have a full generalization to harmonic mappings, because the converse to the lemma is false.
In other words, if $F = H + \overline{G}$ is a convex mapping, the function $f = h + \overline{g}$
with $h(z) = zH'(z)$ and $g(z) = -zG'(z)$ need not be a starlike mapping. In
fact, f need not be univalent. For a counterexample we need look no farther
than the convex mapping

$$L(z) = \operatorname{Re}\{\ell(z)\} + i \operatorname{Im}\{k(z)\}$$

obtained by vertical shearing of the conformal half-plane mapping $\ell(z) = z/(1-z)$ (see Section 3.4.). It has the form $L = H + \overline{G}$, where

$$H(z) = \frac{1}{2}[\ell(z) + k(z)] = \frac{z - \frac{1}{2}z^2}{(1-z)^2}$$

$$G(z) = \frac{1}{2}[\ell(z) - k(z)] = \frac{-\frac{1}{2}z^2}{(1-z)^2}.$$

Thus,

$$h(z) = zH'(z) = \frac{z}{(1-z)^3} \quad \text{and} \quad g(z) = -zG'(z) = \frac{z^2}{(1-z)^3},$$

so that

$$h'(z) = \frac{1+2z}{(1-z)^4} \quad \text{and} \quad g'(z) = \frac{2z+z^2}{(1-z)^4}.$$

The function $f = h + \overline{g}$ therefore has a Jacobian $J(z) = |h'(z)|^2 - |g'(z)|^2$
that changes sign in the unit disk. For instance, $J(0) > 0$ and $J(-\frac{1}{2}) < 0$.
Thus, f is not univalent, by Lewy's theorem (Section 2.2).

The proof of the lemma fails in the converse direction because the Radó–
Kneser–Choquet theorem does not apply to starlike curves.

7

Mapping Problems

7.1. Generalized Riemann Mapping Theorem

According to Riemann's theorem, the unit disk can be mapped conformally onto an arbitrary simply connected domain $\Omega \neq \mathbb{C}$, and the mapping is unique up to precomposition with conformal self-mappings of the disk. It would be desirable to find a suitable analogue for harmonic mappings.

Of course, the existence of a harmonic mapping is not in question, because every conformal mapping is harmonic. The problem is rather an *embarras de richesse*: there are far too many harmonic mappings of the disk onto a given region. This is already evident from the Radó–Kneser–Choquet theorem (Section 3.1) when the target region Ω is convex. The aim is to find some way to classify the mappings and to specify a particular harmonic mapping of $\mathbb{D}$ onto Ω. It may be hoped that some additional data will uniquely determine the mapping function.

At this point the theory of quasiconformal mappings offers a clue. According to a basic theorem of that subject, if μ is a bounded measurable function defined in $\mathbb{D}$ with essential supremum $||\mu||_\infty < 1$, then there is a homeomorphism f of $\mathbb{D}$ onto Ω with complex dilatation $\mu = f_{\bar{z}}/f_z$. In other words, f is a solution to the *Beltrami equation* $f_{\bar{z}} = \mu f_z$. The mapping f is unique up to postcomposition with an arbitrary conformal self-mapping of Ω; this does not change the complex dilatation. Thus, the mapping is formally determined by the requirements that $f(0) = w_0$ and $f_z(0) > 0$, where w_0 is a given point of Ω. In fact, f is quasiconformal because $||\mu||_\infty < 1$. (See Lehto and Virtanen [1] and Ahlfors [1] for careful statements and proofs of this fundamental theorem.)

In view of the theorem just described, one is tempted to believe that the analytic dilatation $\omega = \overline{f_{\bar{z}}}/f_z$ will play a corresponding role in classifying the harmonic mappings onto a given region. Thus, a plausible generalization of the Riemann mapping theorem would run as follows.

Proposed Theorem. *For any given simply connected domain* $\Omega \neq \mathbb{C}$ *and any analytic function* ω *with* $|\omega(z)| < 1$ *in* $\mathbb{D}$, *there exists a sense-preserving harmonic mapping* f *of* $\mathbb{D}$ *onto* Ω *with dilatation* $\omega = \overline{f_{\bar{z}}}/f_z$. *Moreover, such a mapping* f *exists and is unique under the normalization* $f(0) = w_0$ *and* $f_z(0) > 0$, *where* w_0 *is any preassigned point of* Ω.

Riemann's theorem is the case where $\omega(z) \equiv 0$. In fact, it allows us to handle the slightly more general case where $\omega(z)$ is constant. It was remarked in Section 1.2 that if f has dilatation ω, then its affine transform

$$F = \alpha f + \beta \overline{f}, \qquad |\beta| < |\alpha|,$$

has dilatation

$$\overline{F_{\bar{z}}}/F_z = \frac{\overline{\alpha}\omega + \overline{\beta}}{\alpha + \beta\omega}.$$

Hence, if $\omega(z) \equiv c$ for some complex constant c with $|c| < 1$, the choice $\beta = -\alpha\overline{c}$ will make F analytic. In this way it is found that the most general harmonic mapping f with constant dilatation c has the form $f = h + \overline{c}\overline{h}$, where h is a conformal mapping. By inverting the affine mapping $\zeta \mapsto \zeta + \overline{c}\overline{\zeta}$ and appealing to the Riemann mapping theorem, it is possible to find a (unique) conformal mapping h such that f maps $\mathbb{D}$ onto the given domain Ω and has the properties $f(0) = w_0$ and $f_z(0) > 0$. Alternatively, the normalization can be achieved through the simple observation that because the dilatation is constant it is unchanged when the harmonic mapping is precomposed with a conformal self-mapping of the disk.

It turns out that the "proposed theorem" stated above is too ambitious and is false as stated. However, it will be shown in this chapter that the existence assertion is true under certain additional hypotheses or when the conclusion is suitably modified. The uniqueness assertion will be verified for a special class of domains. But before turning to the positive results, it will be instructive to consider some counterexamples.

7.2. Collapsing

The harmonic version of the Riemann mapping theorem, as proposed in the previous section, is not true. Given a simply connected domain $\Omega \subset \mathbb{C}$ and an analytic function ω in the unit disk $\mathbb{D}$ with $|w(z)| < 1$, there need not exist a harmonic mapping of $\mathbb{D}$ onto Ω with dilatation ω.

A counterexample is readily available. As Hengartner and Schober [5] discovered, there is no harmonic mapping of the disk onto itself with $\omega(z) = z$.

The idea of the proof is to show that if f maps $\mathbb{D}$ harmonically *into* itself and has analytic dilatation $\omega(z) = z$, then its image must have area less than π (in fact, less than $\pi/2$), so it cannot fill the disk.

Let $f = h + \bar{g}$ be an arbitrary sense-preserving harmonic mapping of $\mathbb{D}$, and write

$$h(z) = \sum_{n=0}^{\infty} a_n z^n \qquad \text{and} \qquad g(z) = \sum_{n=1}^{\infty} b_n z^n.$$

Then the area of the image $f(\mathbb{D})$ is

$$A = \iint_{\mathbb{D}} J_f(z)\,dx\,dy = \iint_{\mathbb{D}} (|f_z|^2 - |f_{\bar{z}}|^2)\,dx\,dy.$$

$$= \iint_{\mathbb{D}} (|h'(z)|^2 - |g'(z)|^2)\,dx\,dy = \pi \sum_{n=1}^{\infty} n(|a_n|^2 - |b_n|^2)$$

by the Dirichlet formula. Suppose now that f has analytic dilatation $\omega(z) = z$. Then the differential equation $\overline{f_{\bar{z}}} = \omega f_z$ becomes $g'(z) = zh'(z)$, which is equivalent to

$$(n+1)b_{n+1} = na_n, \qquad n = 0, 1, 2, \ldots .$$

In particular, $b_1 = 0$. Thus,

$$A = \pi \sum_{n=1}^{\infty} (n|a_n|^2 - (n+1)|b_{n+1}|^2) = \pi \sum_{n=1}^{\infty} \frac{n}{n+1}|a_n|^2.$$

On the other hand, the integral mean

$$M_2(r, f)^2 = \frac{1}{2\pi} \int_0^{2\pi} |f(re^{i\theta})|^2\,d\theta$$

$$= \frac{1}{2\pi} \int_0^{2\pi} \{|h(re^{i\theta})|^2 + |g(re^{i\theta})|^2\}\,d\theta$$

$$= \sum_{n=0}^{\infty} |a_n|^2 r^{2n} + \sum_{n=1}^{\infty} |b_n|^2 r^{2n}.$$

If $f(\mathbb{D}) \subset \mathbb{D}$, then clearly $M_2(r, f) \leq 1$, so it follows by letting r approach 1 that

$$M_2(1, f)^2 = \sum_{n=0}^{\infty} (|a_n|^2 + |b_{n+1}|^2) \leq 1.$$

If $\omega(z) = z$, this inequality takes the form

$$M_2(1, f)^2 = \sum_{n=0}^{\infty} \left\{ 1 + \left(\frac{n}{n+1} \right)^2 \right\} |a_n|^2 \le 1.$$

But now the trivial inequality

$$2 \left(\frac{n}{n+1} \right) < 1 + \left(\frac{n}{n+1} \right)^2$$

implies that

$$A < \frac{\pi}{2} M_2(1, f)^2 \le \frac{\pi}{2}.$$

In other words, if $f(\mathbb{D}) \subset \mathbb{D}$ and $\omega(z) = z$, then the area of $f(\mathbb{D})$ is less than $\frac{\pi}{2}$. Since the area of $\mathbb{D}$ is π, this shows that f cannot map $\mathbb{D}$ onto $\mathbb{D}$.

Using Hall's sharp version of the Heinz inequality (see Section 4.4), Hengartner and Schober actually improved the estimate to $A \le \frac{\pi}{2} - \frac{27}{32\pi} = 1.302\ldots$. On the other hand, the "extremal function" for the Heinz problem is a harmonic mapping of $\mathbb{D}$ onto an equilateral triangle inscribed in the unit circle, with dilatation $\omega(z) = z$, as shown in Section 4.2. Since the triangle has area $3\sqrt{3}/4 = 1.299\ldots$, one suspects that $A \le 3\sqrt{3}/4$ if $f(\mathbb{D}) \subset \mathbb{D}$ and $\omega(z) = z$, but this has not been proved.

If $\omega(z) = z^2$ and $f(\mathbb{D}) \subset \mathbb{D}$, a calculation similar to the preceding one shows that $A < \pi$ so, again, f cannot map $\mathbb{D}$ onto $\mathbb{D}$. Here we know that there is a harmonic mapping of $\mathbb{D}$ onto a square inscribed in the unit circle, with $\omega(z) = z^2$. The area of the square is 2, so it may be conjectured that $A \le 2$ if $f(\mathbb{D}) \subset \mathbb{D}$ and $\omega(z) = z^2$. In general, it may be conjectured that if $f(\mathbb{D}) \subset \mathbb{D}$ and $\omega(z) = z^{n-2}$ for $n = 3, 4, \ldots$, then $A \le (n/2) \sin(2\pi/n)$, the area of a regular n-gon inscribed in the unit circle (see Section 4.2). This would imply, in particular, that no harmonic mapping of $\mathbb{D}$ onto $\mathbb{D}$ exists with $\omega(z) = z^n$, a fact to be established more generally in Sections 7.3 and 7.4.

Another example is given by the mapping

$$f(z) = z + \frac{1}{n-1} \bar{z}^{n-1}, \qquad n = 3, 4, \ldots.$$

It has been observed (see Section 1.1) that f maps $\mathbb{D}$ onto the interior of
• a hypocycloid of n cusps inscribed in the circle $|w| = 1 + 1/(n-1)$. Thus $[(n-1)/n] f$ maps $\mathbb{D}$ into itself and has dilatation $\omega(z) = z^{n-2}$. It can be shown (*cf.* Section 6.1) that the area A of the image $[(n-1)/n] f(\mathbb{D})$ is

$$A = \frac{(n-1)(n-2)}{n^2} \pi < \frac{n}{2} \sin \frac{2\pi}{n}.$$

The phenomenon illustrated by the example of Hengartner and Schober is known as *collapsing*. It will be seen that collapsing is inherent to the harmonic Riemann mapping problem and that with suitable interpretation it provides a key to positive results.

7.3. Concavity of the Boundary

The example in the previous section exhibits the need for collapsing by displaying an upper bound on the area of the image. However, the phenomenon is better understood from the viewpoint of boundary behavior. If a sense-preserving harmonic mapping has dilatation of unit modulus on some boundary arc, then it sends that arc onto a *concave* arc unless the boundary function is piecewise constant or stationary. A theorem to this effect will be formulated presently. As a consequence, a harmonic mapping of the disk onto a strictly convex domain cannot have dilatation of unit modulus on the boundary unless its boundary function is stationary. Recall that every harmonic mapping onto a strictly convex domain does extend continuously to the boundary (see Section 3.3).

The concavity property is illustrated by many of the specific harmonic mappings considered in earlier chapters. For example, the mapping $f(z) = z + (1/n)\bar{z}^n$ has dilatation $\omega(z) = z^{n-1}$ and it maps the disk onto the region inside a hypocycloid of $n + 1$ cusps (see Figure 1.1). Here $|\omega(z)| = 1$ on the whole boundary, yet the function has a homeomorphic extension to the closure of the disk. However, the boundary of the image is nowhere convex. Note that the cusps correspond to the critical points of the boundary function, the points $\theta = 2k\pi/(n + 1)$ where $d/d\theta\{f(e^{i\theta})\} = 0$.

Further examples come from the shear construction. For instance, if the identity function is sheared with dilatation $\omega(z) = z$, the resulting harmonic mapping $f(z) = -\bar{z} - 2\log|1 - z|$ sends the unit disk onto a "three-cornered hat" with concave boundary arcs, as depicted in Figure 3.1. Here the mapping f can be shown to have a homeomorphic extension to the closed disk, with respect to the spherical metric. Similar remarks apply to the harmonic mapping produced by shearing the identity with dilatation $\omega(z) = z^2$. Its image is shown in Figure 3.2.

Another typical class of examples is discussed in Chapter 10 in connection with Scherk's saddle-tower minimal surface (see Figure 10.2).

On the other hand, the shears of the horizontal strip mapping $s(z) = \frac{1}{2}\log\frac{1+z}{1-z}$ with dilatations z and z^2 are mappings onto convex regions, a half-strip and a full strip, respectively; and their boundary functions are piecewise constant. The images of these harmonic mappings are displayed in Figures 3.4 and 3.5.

The following theorem helps to explain the concavity of image persistently apparent in harmonic mappings with dilatations of unit modulus on the boundary.

Theorem. *Let f be a sense-preserving harmonic mapping of $\mathbb{D}$ onto a domain Ω. Suppose that f has a C^1 extension to some open arc $I \subset \mathbb{T}$ that it maps univalently onto a convex arc $\gamma \subset \partial\Omega$. Let s denote arclength along γ as a function of θ for $e^{i\theta} \in I$. Suppose $ds/d\theta \neq 0$ at some point $\zeta = e^{i\alpha} \in I$. Then the dilatation of f has a continuous extension to a subarc of I containing ζ, and $|\omega(\zeta)| < 1$.*

Corollary. *Let f be a sense-preserving harmonic mapping of $\mathbb{D}$ onto a convex domain Ω and suppose f has a C^1 extension to a homeomorphism of $\overline{\mathbb{D}}$ onto $\overline{\Omega}$ with $ds/d\theta > 0$ at every boundary point. Then f is quasiconformal.*

Deduction of Corollary. The dilatation $\omega = \overline{f_{\bar{z}}}/f_z$ has the property $|\omega(z)| < 1$ in $\mathbb{D}$, and it has a continuous extension to $\overline{\mathbb{D}}$ which, by the theorem, still satisfies $|\omega(z)| < 1$ everywhere on $\mathbb{T}$. Thus, $|\omega(z)| \leq k$ for some constant $k < 1$, and f is quasiconformal in $\mathbb{D}$.

The theorem says that on an arc of the circle where $|\omega(e^{i\theta})| = 1$, the boundary function cannot move smoothly with positive speed along a convex boundary arc; it must either remain stationary, jump, or move along a strictly concave boundary arc. This principle is implicit in work of Hengartner and Schober [4], further refined by Bshouty and Hengartner [4], which will be discussed in the next section. The relatively simple approach to be followed here is due to Duren and Khavinson [1]. The corollary generalizes and slightly improves a theorem by Olli Martio [1] for harmonic self-mappings of the disk. The result can be viewed as a partial converse of the well-known theorem (a strong form of the Carathéodory extension theorem) that any quasiconformal mapping of one Jordan domain onto another extends to a homeomorphism of the closures.

The theorem will be proved with the aid of Hopf's lemma, which is included here for the sake of completeness.

Hopf's Lemma. *Let $D \subset \mathbb{C}$ be a Jordan domain with smooth boundary. Let u be a nonconstant harmonic function in D that has a smooth extension to $\overline{D}$. If $u(z)$ has a local minimum at some point $\zeta \in \partial D$, then its inner normal derivative is strictly positive at that point: $\frac{\partial u}{\partial n}(\zeta) > 0$.*

Proof of Lemma. Suppose without loss of generality that $u(\zeta) = 0$ and $u(z) > 0$ for points in D sufficiently near ζ. Then $u(z) > 0$ in some disk $\Delta \subset D$ of radius r whose boundary is tangent to ∂D at ζ. Suppose for convenience, after rotation and translation, that Δ is centered at the origin and $\zeta = r$. Then, by Harnack's inequality (see Section 1.4),

$$u(x) \geq u(0)\frac{r - x}{r + x}, \qquad 0 < x < r.$$

Thus,

$$\frac{u(x) - u(r)}{r - x} \geq \frac{u(0)}{r + x}, \qquad 0 < x < r.$$

Now let x tend to r to conclude that

$$\frac{\partial u}{\partial n}(r) = -\frac{\partial u}{\partial x}(r) \geq \frac{u(0)}{2r} > 0. \qquad \blacksquare$$

Proof of Theorem. Preceding f by a conformal mapping, we may replace the unit disk by a domain D in the upper half-plane with a boundary arc I on the real axis, containing the origin. After rotation and translation, we may assume that the image domain Ω lies in the upper half-plane and is tangent to the real axis at the origin and that $f(0) = 0$. The hypothesis is then that f has a smooth homeomorphic extension to I and that $f(I)$ is a convex boundary arc of Ω. With the notation $w = f(z)$, where $z = x + iy$ and $w = u + iv$, it is clear that $v_x(0) = 0$, since v has a local minimum at 0. Thus, the hypothesis of positive speed allows the assumption that $u_x(0) > 0$, and the Jacobian of f at the origin is

$$J(0) = u_x(0)v_y(0) - u_y(0)v_x(0) = u_x(0)v_y(0).$$

Observe now that v is a positive harmonic function in D that attains its minimum value of 0 at the origin. Hence, by Hopf's lemma, it follows that $v_y(0) > 0$, since that is the inner normal derivative of v at the origin. Thus, $J(0) > 0$ or, equivalently, $|h'(0)| > |g'(0)|$, which shows that the dilatation $\omega = g'/h'$ is continuous up to a subarc of I containing the origin and $|\omega(0)| < 1$. This proves the theorem. $\blacksquare$

It may be noted that the proof extends with little change to sense-preserving harmonic functions, not necessarily univalent. Duren and Khavinson [1] supplied an example to show that the hypothesis $ds/d\theta > 0$ is essential.

For harmonic mappings whose range is convex in one direction, the representation formula arising from the shear construction (see Section 3.4) can be used to demonstrate the concavity property by direct calculation. Details were given by (Greiner [1, 2]).

7.4. Angles at Corners

It was found in Section 4.3 that certain harmonic mappings of the disk onto convex polygonal regions, generated by piecewise constant boundary functions, always have finite Blaschke products as dilatations. We are now ready to establish a kind of converse; namely, that if a harmonic mapping sends the disk "onto" the interior of a convex curve (in a generalized sense to be made precise) and if its dilatation is a finite Blaschke product, then its range is actually a polygonal region inscribed in that curve. This shows in particular that collapsing *must* occur whenever the prescribed dilatation is a finite Blaschke product. The result is an application of a "corner condition" found by Hengartner and Schober, which sheds further light on the mechanism of collapsing and shows that the proposed generalization of the Riemann mapping theorem meets obstructions more fundamental than limitations on area of range.

We begin with a general theorem from Hengartner and Schober [4]. Recall first that a bounded harmonic function in the disk has a radial limit $\hat{f}(e^{i\theta}) = \lim_{r \to 1} f(re^{i\theta})$ almost everywhere (see, for instance, Duren [1], Ch. 1).

Theorem 1. *Let ω be a finite Blaschke product and let f be a nonconstant bounded solution of the equation $\overline{f_{\bar{z}}} = \omega f_z$ in $\mathbb{D}$. Then the boundary function $\hat{f}(e^{i\theta}) - \overline{\omega(e^{i\theta})} \, \overline{\hat{f}(e^{i\theta})}$ agrees almost everywhere with an absolutely continuous function $\varphi(\theta)$, and*

$$\varphi'(\theta) = i e^{-i\theta} \overline{\omega'(e^{i\theta})} \overline{\hat{f}(e^{i\theta})} \qquad \text{a.e.}$$

If it could be shown that in the canonical representation $f = h + \overline{g}$ the derivatives h' and g' belong to the Hardy space H^1, the conclusions of the theorem would follow from known consequences of the F. and M. Riesz theorem. Specifically, it would then follow that h and g are continuous in $\overline{\mathbb{D}}$, that their boundary functions are absolutely continuous, and that the derivatives of those boundary functions agree essentially with the radial limits of h' and g' (see Duren [1], p. 42). Theorem 1 could then be verified by direct calculation.

As may be expected, the proof proceeds along similar lines. The basic tool is the F. and M. Riesz theorem, recorded here for reference (for a proof, see Duren [1], Ch. 3).

Theorem (F. and M. Riesz). *Let $\mu(t)$ be a complex-valued function of bounded variation on $[0, 2\pi]$, with the property*

$$\int_0^{2\pi} e^{int} \, d\mu(t) = 0, \qquad n = 1, 2, \ldots.$$

Then $\mu(t)$ agrees almost everywhere with an absolutely continuous function.

Proof of Theorem. Consider the level-set Γ_r defined by $|\omega(z)| = r$ for $r < 1$. If r is sufficiently close to 1, it can be seen that Γ_r is a simple closed curve in $\mathbb{D}$ with a parametrization

$$z = z_r(t) = \rho_r(t)e^{it}, \qquad 0 \le t \le 2\pi,$$

where ρ_r is continuous and $0 < \rho_r(t) < 1$. Note that $\rho_r(t) \to 1$ uniformly in $[0, 2\pi]$ as $r \to 1$, and the total variation

$$\int_{\Gamma_r} |dz| = \int_0^{2\pi} |dz_r(t)| \to 2\pi.$$

Now define the complex measure $d\mu_r = df - \overline{\omega} \, \overline{df}$ on Γ_r, where $df = f_z \, dz + f_{\bar{z}} \, \overline{dz}$. Since f satisfies the equation $\overline{f_{\bar{z}}} = \omega f_z$, a simple calculation shows that

$$d\mu_r = (1 - |\omega|^2) f_z \, dz = (1 - r^2) f_z \, dz.$$

But f is harmonic, so f_z is analytic and

$$\int_{\Gamma_r} z^k \, d\mu_r = (1 - r^2) \int_{\Gamma_r} z^k f_z(z) \, dz = 0, \qquad k = 0, 1, 2, \ldots,$$

by Cauchy's theorem.

Next let $\mu_r(t)$ denote the measure of the subarc of Γ_r from $z_r(0)$ to $z_r(t)$. Then $\mu_r(0) = 0$ and

$$d\mu_r(t) = (1 - |\omega(z_r(t))|^2) f_z(z_r(t)) \, dz_r(t).$$

But the Poisson formula gives the inequality

$$|f_z(z)| \le \|f\|_\infty \frac{1}{2\pi} \int_0^{2\pi} |e^{it} - z|^{-2} \, dt = \|f\|_\infty (1 - |z|^2)^{-1},$$

where $\|f\|_\infty$ is the supremum of $|f(z)|$ in $\mathbb{D}$. Moreover, the lemma in Section 4.3 shows that

$$1 - |\omega(z)|^2 \le M(1 - |z|^2), \qquad z \in \mathbb{D},$$

where M is a constant. A combination of the two estimates gives $|d\mu_r(t)| \leq M\|f\|_\infty |dz_r(t)|$, showing that μ_r is of uniformly bounded variation for r near 1. It now follows from the Helly selection theorem (see, *e.g.*, Duren [1], p. 3) that some subsequence $\{\mu_{r_j}(t)\}$ converges almost everywhere to a function $\mu(t)$, and

$$\int_0^{2\pi} e^{ikt}\, d\mu(t) = \lim_{j\to\infty} \int_0^{2\pi} e^{ikt}\, d\mu_{r_j}(t) = \lim_{j\to\infty} \int_{\Gamma_r} z^k\, d\mu_{r_j} = 0$$

for $k = 0, 1, 2, \ldots$, since $\rho_r(t) \to 1$ uniformly and we have already observed that $\int_{\Gamma_r} z^k\, d\mu_r = 0$. Thus, $\mu(t)$ is absolutely continuous by the F. and M. Riesz theorem.

On the other hand, an integration by parts based on the original expression $d\mu_r = df - \overline{\omega}\,\overline{df}$ gives

$$\mu_r(t) = f(z_r(t)) - f(z_r(0)) - \overline{\omega(z_r(t))\, f(z_r(t))}$$
$$+ \overline{\omega(z_r(0))\, f(z_r(0))} + \int_0^t \overline{f(z_r(\tau))\, d\omega(z_r(\tau))}.$$

Letting $r \to 1$, we conclude that

$$\mu(t) = \varphi(t) - i \int_0^t \overline{\hat{f}(e^{i\tau})} e^{-i\tau} \overline{\omega'(e^{i\tau})}\, d\tau + C,$$

where C is a constant and

$$\varphi(t) = \hat{f}(e^{it}) - \overline{\omega(e^{it})}\, \overline{\hat{f}(e^{it})} \qquad \text{a.e.}$$

In particular, $\varphi(t)$ is absolutely continuous and

$$\varphi'(t) = \mu'(t) + i e^{-it} \overline{\omega'(e^{it})}\, \overline{\hat{f}(e^{it})} \qquad \text{a.e.}$$

The final step is to show that $\mu'(t) = 0$ a.e. Indeed, it is easily seen that $\overline{\varphi} = -\omega\varphi$, and so

$$\overline{\mu'(t)} = -\frac{d}{dt}\{\omega(e^{it})\varphi(t)\} + \hat{f}(e^{it})\frac{d}{dt}\omega(e^{it})$$
$$= -\omega(e^{it})\varphi'(t) + [\hat{f}(e^{it}) - \varphi(t)]\frac{d}{dt}\omega(e^{it})$$
$$= -\omega(e^{it})\varphi'(t) + \overline{\hat{f}(e^{it})}\,\overline{\omega(e^{it})}\frac{d}{dt}\omega(e^{it})$$
$$= -\omega(e^{it})\mu'(t) \qquad \text{a.e.},$$

since $|\omega(e^{it})| = 1$. It follows that

$$\int_0^{2\pi} e^{ikt}\overline{\mu'(t)}\,dt = -\int_0^{2\pi} e^{ikt}\omega(e^{it})\mu'(t)\,dt = 0$$

for $k = 0, 1, 2, \ldots$, because $\omega(z)$ can be approximated by polynomials uniformly on $\mathbb{T}$ and we have already seen that

$$\int_0^{2\pi} e^{ikt}\mu'(t)\,dt = 0 \qquad \text{for } k = 0, 1, 2, \ldots.$$

This shows that $\mu'(t) = 0$ a.e., and the proof is complete. ∎

Corollary. *Under the hypotheses of the theorem, Im $\{\sqrt{\omega(e^{i\theta})}\,\hat{f}(e^{i\theta})\}$ agrees locally almost everywhere with an absolutely continuous function, and*

$$\operatorname*{ess\,lim}_{t\to 0} \operatorname{Im}\left\{ \sqrt{\omega(e^{i\theta})}\,\frac{1}{t}\,[\hat{f}(e^{i(\theta+t)}) - \hat{f}(e^{i\theta})] \right\} = 0 \qquad a.e.$$

Here either branch of the square root may be chosen.

Proof. To verify the first statement, observe that

$$\sqrt{\omega(e^{i\theta})}\,\varphi(\theta) = 2i\,\operatorname{Im}\left\{ \sqrt{\omega(e^{i\theta})}\,\hat{f}(e^{i\theta}) \right\} \qquad \text{a.e.,}$$

where φ is the absolutely continuous function defined in the theorem. Thus, to verify the second statement, one can adapt the standard proof of the product rule for derivatives (in elementary calculus) to see that it is equivalent to show

$$\frac{d}{d\theta}\left\{ \sqrt{\omega(e^{i\theta})}\,\varphi(\theta) \right\} = 2i\operatorname{Im}\left\{ \hat{f}(e^{i\theta})\frac{d}{d\theta}\sqrt{\omega(e^{i\theta})} \right\} \qquad \text{a.e.}$$

But this follows by straightforward calculation after writing

$$\frac{d}{d\theta}\left\{ \sqrt{\omega(e^{i\theta})}\,\varphi(\theta) \right\} = \frac{d}{d\theta}\left\{ \sqrt{\omega(e^{i\theta})} \right\}\varphi(\theta) + \sqrt{\omega(e^{i\theta})}\,\varphi'(\theta)$$

and substituting the formulas for φ and φ' given in Theorem 1. ∎

In order to understand the implications of the corollary, it is useful to reexamine the mapping of the disk onto an inscribed regular n-gon as constructed in Section 4.2. With $\alpha = e^{2\pi i/n}$, the mapping function f was defined as the Poisson integral of the piecewise continuous boundary function

$$\hat{f}(e^{i\theta}) = \alpha^k, \qquad (2k-1)\pi/n < \theta < (2k+1)\pi/n,$$

for $k = 1, 2, \ldots, n$. The dilatation of f was found to be $\omega(z) = z^{n-2}$. According to the corollary just proved,

$$\sqrt{\omega(e^{i\theta})}\,[\hat{f}(e^{i\theta+}) - \hat{f}(e^{i\theta-})]$$

is purely real everywhere, where

$$\hat{f}(e^{i\theta-}) = \lim_{t \nearrow \theta} \hat{f}(e^{it}) \qquad \text{and} \qquad \hat{f}(e^{i\theta+}) = \lim_{t \searrow \theta} \hat{f}(e^{it})$$

denote the one-sided limits of $\hat{f}(e^{i\theta})$. At the jump points of the boundary function, this can be viewed as a corner condition, specifying the angle at each vertex. To be more precise, note that $\hat{f}(e^{i\theta})$ jumps at the points $e^{i\theta} = \beta^{2k-1}$, $k = 1, 2, \ldots, n$, where $\beta = e^{\pi i/n}$; and at these points

$$\hat{f}(e^{i\theta+}) - \hat{f}(e^{i\theta-}) = \alpha^k - \alpha^{k-1} = \alpha^{k-1}(\alpha - 1),$$

while

$$\sqrt{\omega(e^{i\theta})} = \pm\, i\alpha^{\frac{1}{2}-k}$$

since $\alpha^n = 1$. Thus, for $e^{i\theta} = \beta^{2k-1}$, a short calculation gives the value

$$\sqrt{\omega(e^{i\theta})}\,[\hat{f}(e^{i\theta+}) - \hat{f}(e^{i\theta-})] = \pm 2 \sin\frac{\pi}{n},$$

which is indeed purely real.

The next theorem exhibits the collapsing phenomenon in a rather general setting. It shows in particular that no harmonic mapping whose dilatation is a finite Blaschke product can send the disk onto a strictly convex region, such as another disk.

Theorem 2. *Let ω be a finite Blaschke product of degree n and let f be a harmonic mapping of $\mathbb{D}$ into a bounded convex domain Ω satisfying $\overline{f_{\bar{z}}} = \omega f_z$. Suppose also that the radial limit $\hat{f}(e^{i\theta})$ lies almost everywhere on the boundary $\partial\Omega$. Then aside from a set $E \subset \mathbb{T}$ of exactly $n+2$ points, the function f has a continuous extension to $\overline{\mathbb{D}}$, and the boundary function $\hat{f}(e^{i\theta})$ is constant on each arc of $\mathbb{T} \setminus E$. Furthermore, f maps $\mathbb{D}$ onto the interior of a polygon with at most $n+2$ vertices, all on $\partial\Omega$. If Ω is strictly convex, the polygon has precisely $n+2$ vertices.*

Proof. According to the theorem in Section 3.3, the function f has an unrestricted limit

$$\hat{f}(e^{it}) = \lim_{z \to e^{it}} f(z)$$

at each point $e^{it} \in \mathbb{T} \setminus E$, where the exceptional set E is at most countable. At each point $e^{i\theta} \in \mathbb{T}$ the one-sided limits $\hat{f}(e^{i\theta-})$ and $\hat{f}(e^{i\theta+})$ exist; they are equal for $e^{i\theta} \in \mathbb{T} \setminus E$ and unequal for $e^{i\theta} \in E$.

We shall show first that under the present hypotheses the exceptional set E is finite and contains no more than $n+2$ points. By the Corollary to Theorem 1, the quantity

$$\sqrt{\omega(e^{i\theta})} \, [\hat{f}(e^{i\theta+}) - \hat{f}(e^{i\theta-})]$$

is real everywhere on E. Thus, for $e^{i\theta} \in E$, the function

$$\psi(\theta) = \frac{1}{2}\arg\{\omega(e^{i\theta})\} + \arg\{\hat{f}(e^{i\theta+}) - \hat{f}(e^{i\theta-})\}$$

is an integer multiple of π. But it follows from the univalence of f that $\arg\{\hat{f}(e^{i\theta+}) - \hat{f}(e^{i\theta-})\}$ is a nondecreasing function of θ for $e^{i\theta} \in E$, with net increase of 2π over the interval $0 \le \theta < 2\pi$. Since ω is a Blaschke product of degree n, $\arg\{\omega(e^{i\theta})\}$ increases by $2n\pi$ over the full circle. Therefore, $\psi(\theta)$ is a strictly increasing function with total increase less than $(n+2)\pi$ over E, regarded as a subset of the interval $[0, 2\pi)$. As a consequence, there can be at most $n+2$ points in E where $\psi(\theta)$ is an integer multiple of π. But we have shown that $\psi(\theta)$ is an integer multiple of π for every point $e^{i\theta}$ in E, so E contains at most $n+2$ points.

The next step is to show that $\hat{f}(e^{i\theta})$ is constant on each interval of the complementary set $\mathbb{T} \setminus E$. If not, some interval I of $\mathbb{T} \setminus E$ must contain infinitely many points $e^{i\theta_j}$ for which $\hat{f}(e^{i\theta})$ is nonconstant on the interval $[\theta_j, \theta_j + \varepsilon_j)$ for each $\varepsilon_j > 0$, because $\hat{f}(e^{i\theta})$ is continuous and sense-preserving on I. Since $\arg\{\hat{f}(e^{i\theta}) - \hat{f}(e^{i\theta_j})\}$ is bounded and nondecreasing, the limits

$$\sigma_j = \lim_{\theta \searrow \theta_j} \arg\{\hat{f}(e^{i\theta}) - \hat{f}(e^{i\theta_j})\}$$

exist for each j and are monotonic in the sense that $\sigma_j > \sigma_k$ if $\theta_j > \theta_k$. By the corollary to Theorem 1, $\psi(\theta_j) = \frac{1}{2}\tau_j + \sigma_j$ must always be an integer multiple of π, where $\tau_j = \arg\{\omega(e^{i\theta_j})\}$. But $\psi(\theta)$ is strictly increasing and bounded over I, so it cannot assume integer multiples of π for infinitely many points $e^{i\theta_j}$. This contradiction shows that $\hat{f}(e^{i\theta})$ is constant on I and, hence, on each component of $\mathbb{T} \setminus E$.

Finally, $f(z)$ is the Poisson integral of its piecewise constant boundary function $\hat{f}(e^{i\theta})$. Since its dilatation ω is a finite Blaschke product of degree n, it follows from the result of Sheil-Small, as developed in Section 4.3, that $\hat{f}(e^{i\theta})$ has exactly $n+2$ jump discontinuities. Thus f maps $\mathbb{D}$ onto the interior of a convex polygon with at most $n+2$ vertices, all situated on $\partial\Omega$. There will be exactly $n+2$ vertices unless three consecutive values of $\hat{f}(e^{i\theta})$ are collinear, creating an "invisible" vertex. This cannot happen if Ω is strictly convex, or equivalently if its boundary contains no line segments, so the polygon has exactly $n + 2$ (proper) vertices in this case. The proof of Theorem 2 is complete.   ∎

The question arises whether *every* finite Blaschke product can occur as the dilatation of some harmonic mapping of the disk onto a convex polygon piecewise constant on the boundary. According to the result in Section 4.3, each such mapping gives rise to a finite Blaschke product as its dilatation. Conversely, Theorem 2 says that if it is known *a priori* that a finite Blaschke product ω is the dilatation of a harmonic mapping f "onto" a given convex region, then in fact the range of f is a convex polygon and its boundary function is piecewise constant. If ω has the simple form

$$\omega(z) = \gamma \left[\frac{z - z_1}{1 - \overline{z_1} z} \right]^n,$$

where γ is a unimodular constant, $z_1 \in \mathbb{D}$, and n is a positive integer, an associated harmonic mapping can be constructed explicitly. According to the calculations in Section 4.2, the dilatation γz^n is realized for a suitable harmonic mapping f onto a regular $(n + 2)$-gon inscribed in the unit circle. If f is preceded by the conformal self-map of the disk

$$\varphi(z) = \frac{z - z_1}{1 - \overline{z_1} z},$$

the composition $f \circ \varphi$ has the desired dilatation and again maps the disk harmonically onto the regular $(n + 2)$-gon, with piecewise constant boundary values. (See Section 1.2 for the behavior of dilatation under composition.)

For more general finite Blaschke products ω, the issue is not so clear. If a harmonic mapping with piecewise constant boundary function is to have a prescribed dilatation ω, its values must obey the corner condition implicit in the Corollary to Theorem 1. The following theorem of Hengartner and Schober [4] shows that at least under an additional hypothesis the resulting harmonic extension will have the prescribed dilatation.

Theorem 3. *Let ω be a finite Blaschke product of degree n. Let Ω be a convex polygon with $n + 2$ distinct vertices $\alpha_1, \alpha_2, \dots , \alpha_{n+2}$ taken in counterclockwise order around the boundary. Define a partition $\{e^{it_k}\}$ of the unit circle by choosing numbers $t_0 < t_1 < \cdots < t_{n+2} = t_0 + 2\pi$, and define the* *step function* $\varphi(e^{it}) = \alpha_k$ *for* $t_{k-1} < t < t_k, k = 1, 2, \ldots, n+2$. *Then the harmonic extension*

$$f(z) = \frac{1}{2\pi} \int_0^{2\pi} \frac{1 - |z|^2}{|e^{it} - z|^2} \varphi(e^{it}) \, dt$$

is univalent and maps the unit disk onto Ω. *If* φ *satisfies the corner conditions*

$$(\star) \qquad \operatorname{Im}\left\{ \sqrt{\omega(e^{it_k})}(\alpha_{k+1} - \alpha_k) \right\} = 0, \qquad k = 0, 1, \ldots, n+1,$$

and if all singularities of the function $z[\overline{f_{\bar{z}}(z)}]^2/\omega(z)$ *are removable, then* f *has dilatation* ω.

It should be observed that the extra condition on removable singularities is necessary, in stronger form, if f is to have dilatation ω. For if f is a sense-preserving harmonic mapping with $\overline{f_{\bar{z}}} = \omega f_z$, then, since f_z is nonvanishing, the zeros of ω are precisely those of $\overline{f_{\bar{z}}}$, multiplicities counted. The hypothesis imposed in the theorem requires $\overline{f_{\bar{z}}}$ to vanish wherever ω does (unless ω has a simple zero at the origin), but the order of the zero of $\overline{f_{\bar{z}}}$ is required to be only about half that of ω.

Proof of Theorem 3. As in Section 4.3, the calculation of f_z and $f_{\bar{z}}$ is facilitated by writing the Poisson kernel as the average of $(e^{it} + z)/(e^{it} - z)$ and its complex conjugate. Computing the derivatives and integrating by parts, we find

$$\pi i z[\overline{f_{\bar{z}}(z)} - \omega(z) f_z(z)] = \int_0^{2\pi} \frac{e^{it} + z}{e^{it} - z} \{\overline{d\varphi(e^{it})} - \omega(z) \, d\varphi(e^{it})\}$$

$$= \sum_{k=1}^{n+2} \frac{e^{it_k} + z}{e^{it_k} - z} \{(\overline{\alpha_{k+1}} - \overline{\alpha_k}) - \omega(z)(\alpha_{k+1} - \alpha_k)\},$$

where $\alpha_{k+3} = \alpha_1$. Squaring both sides and dividing by $\omega(z)$, we arrive at the formula

$$\frac{\pi^2 z^2}{\omega(z)} [\overline{f_{\bar{z}}(z)} - \omega(z) f_z(z)]^2$$

$$= -\left[\sum_{k=1}^{n+2} \frac{e^{it_k} + z}{e^{it_k} - z} \left\{ \sqrt{\omega(z)}(\alpha_{k+1} - \alpha_k) - \frac{1}{\sqrt{\omega(z)}}(\overline{\alpha_{k+1}} - \overline{\alpha_k}) \right\} \right]^2.$$

Now the function on the left-hand side of this equation is analytic in $\mathbb{D}$ in view of the hypothesis that the singularities of $z[\overline{f_{\bar{z}}(z)}]^2/\omega(z)$ are removable. The apparent simple poles of the right-hand side at the points $z = e^{it_k}$ are removed by the corner conditions $(\star)$ imposed on φ. Thus, the function on the left-hand side has an analytic extension to $\overline{\mathbb{D}}$. Furthermore, the right-hand side is real on $\mathbb{T}$, since $|\omega(e^{i\theta})| = 1$ and

$$\frac{e^{it_k} + e^{i\theta}}{e^{it_k} - e^{i\theta}} = i \, \cot \frac{1}{2}(\theta - t_k)$$

is imaginary, so that each term of the sum is real for $z = e^{i\theta}$. As a consequence, the left-hand side is identically constant. But it vanishes at the origin, so $\overline{f_{\bar{z}}(z)} - \omega(z)f_z(z) = 0$ everywhere in $\mathbb{D}$. In other words, ω is the dilatation of f. ∎

7.5. Existence Theorems

The collapsing phenomenon shows that the generalized Riemann mapping theorem, proposed at the beginning of this chapter, cannot be true as stated. For a prescribed simply connected domain $\Omega \neq \mathbb{C}$ and a prescribed analytic function ω with $|\omega(z)| < 1$, there may not exist a harmonic mapping f of $\mathbb{D}$ onto Ω with dilatation $\overline{f_{\bar{z}}}/f_z = \omega$, but in principle the theorem is correct. It can be rescued in two ways, either by imposing additional restrictions on the dilatation ω or on the target region Ω, or by reinterpreting the conclusion that f maps the disk "onto" Ω. Hengartner and Schober [5] formulated theorems along each of those lines, as follows.

Theorem 1. *Let Ω be a bounded simply connected domain whose boundary is an analytic Jordan curve and let w_0 be an arbitrary point in Ω. Let ω be analytic and satisfy $|\omega(z)| \leq k$ in $\mathbb{D}$ for some constant $k < 1$. Then there exists a harmonic mapping f of $\mathbb{D}$ onto Ω with dilatation ω having the additional properties $f(0) = w_0$ and $f_z(0) > 0$.*

Theorem 2. *Let Ω be a bounded simply connected domain whose boundary is a Jordan curve Γ, and let w_0 be an arbitrary point in Ω. Let ω be analytic and satisfy $|\omega(z)| < 1$ in $\mathbb{D}$. Then there is a harmonic mapping f of $\mathbb{D}$ into Ω with dilatation ω, satisfying $f(0) = w_0$ and $f_z(0) > 0$, whose radial limits $\hat{f}(e^{i\theta}) = \lim_{r \to 1} f(re^{i\theta})$ belong to Γ for almost every angle θ.*

Theorem 1 is the analogue of a standard theorem in the theory of quasiconformal mappings (see Lehto and Virtanen [1] or Ahlfors [1]), where a measurable function μ is prescribed with $\|\mu\|_\infty < 1$ and a quasiconformal mapping f is found with *first* complex dilatation $f_{\bar{z}}/f_z = \mu$. The corresponding result for the second complex dilatation $\omega = \overline{f_{\bar{z}}}/f_z$ is less well known and is more difficult to prove, but it has been available in some form since the

1940s. A proof based on integral operators and fixed-point theorems was outlined by Hengartner and Schober [5] with reference to Wendland [1]; see also Renelt [1]. Further details will be deferred to the next section.

Theorem 2 is designed to take account of collapsing. Consider, for instance, a harmonic function with piecewise constant boundary values as constructed in Sections 4.2 and 4.3. Such a function maps the disk onto a convex polygon, but its radial limits lie at the vertices. Thus, in the generalized sense of Theorem 2, it will map the disk "onto" the interior of any Jordan curve surrounding the polygon and containing all of the vertices. More specifically, there is no harmonic mapping of the unit disk onto itself with analytic dilatation $\omega(z) = z$, but there is such a mapping onto an equilateral triangle inscribed in the unit circle, and it maps the disk onto itself in the more general sense.

Theorem 2 is due to Hengartner and Schober. They deduced it from Theorem 1 by appeal to the following theorem, which is best viewed as a replacement for the Carathéodory convergence theorem in the context of harmonic mappings.

Theorem 3. *Let Ω be a bounded simply connected domain whose boundary is a Jordan curve Γ, and let w_0 be an arbitrary point in Ω. Let φ be the conformal mapping of $\mathbb{D}$ onto Ω normalized by $\varphi(0) = w_0$ and $\varphi'(0) > 0$. Choose an increasing sequence $\{r_n\}$ with $0 < r_n < 1$ and $\lim_{r \to 1} r_n = 1$. Let $\mathbb{D}_n$ be the subdisk $\{z : |z| < r_n\}$ and let $\Omega_n = \varphi(\mathbb{D}_n)$, so that $w_0 \in \Omega_n$ for all n. Let $\{\omega_n\}$ be a sequence of analytic functions with $|\omega_n(z)| < 1$ in $\mathbb{D}$ and $\sup_n |\omega_n(0)| < 1$. Suppose there are harmonic mappings f_n of $\mathbb{D}$ onto Ω_n, extending to homeomorphisms of $\overline{\mathbb{D}}$ onto $\overline{\Omega_n}$, with dilatations ω_n and the properties $f_n(0) = 0$ and $(f_n)_z(0) > 0$. Then there is a subsequence $\{n_j\}$ such that $\omega_{n_j}(z) \to \omega(z)$ and $f_{n_j}(z) \to f(z)$ uniformly on compact subsets of $\mathbb{D}$, and $|\omega(z)| < 1$ in $\mathbb{D}$. Furthermore, f is a harmonic mapping of $\mathbb{D}$ into Ω with dilatation ω and the properties $f(0) = w_0$ and $f_z(0) > 0$, whose radial limits $\hat{f}(e^{i\theta}) = \lim_{r \to 1} f(re^{i\theta})$ lie on Γ for almost every angle θ.*

In other words, as the domains Ω_n expand to fill Ω, the corresponding harmonic mappings f_n converge to a function f that maps the disk onto Ω in the generalized sense.

Proof of Theorem 2. An easy argument derives the main mapping theorem from Theorems 1 and 3. As in the statement of Theorem 3, let φ be the normalized conformal mapping of $\mathbb{D}$ onto Ω, choose an increasing sequence $\{r_n\}$ tending to 1, and define $\Omega_n = \varphi(\mathbb{D}_n)$, where $\mathbb{D}_n$ is the disk $|z| < r_n$.

Let $\omega_n(z) = \omega(r_n z)$, so that $\omega_n(0) = \omega(0) < 1$ and $|\omega_n(z)| \leq k_n < 1$ in $\mathbb{D}$. By Theorem 1, there exists a harmonic mapping f_n of $\mathbb{D}$ onto Ω_n with dilatation ω_n, which is normalized so that $f_n(0) = w_0$ and $(f_n)_z(0) > 0$. By the generalization of the Carathéodory extension theorem to quasiconformal mappings (see Lehto and Virtanen [1], p. 42), f_n extends to a homeomorphism of $\overline{\mathbb{D}_n}$ onto $\overline{\Omega_n}$. This puts us in position to apply Theorem 3. It allows us to assert that a subsequence $\{f_{n_j}\}$ converges uniformly on compact subsets of $\mathbb{D}$ to a harmonic mapping f of $\mathbb{D}$ into Ω, with dilatation ω and normalization $f(0) = w_0$, $f_z(0) > 0$. We can also conclude from Theorem 3 that f maps $\mathbb{D}$ onto Ω in the weak sense; its radial limits lie almost everywhere on Γ. ∎

Proof of Theorem 3. The family of functions f_n is uniformly bounded in D, so it follows from a variant of Montel's theorem for harmonic functions (*cf.* Ahlfors [3], p. 224) that a subsequence $\{f_{n_j}\}$ converges locally uniformly to a function f harmonic in $\mathbb{D}$. To see that f is not constant, consider the normalized functions

$$g_n(z) = [f_n(z) - w_0]/(f_n)_z(0).$$

Since $g_n \in S_H$, a result of Clunie and Sheil-Small (Theorem 1 in Section 6.2) shows that g_n omits a value of modulus less than 2 (actually less than 1.72). Thus, the distance δ_n from w_0 to the boundary of Ω_n is greater than $2(f_n)_z(0)$, so that

$$(f_n)_z(0) > \delta_n/2 \geq \delta > 0,$$

where $\delta = \delta_1/2$. Now let n tend to infinity through the sequence $\{n_j\}$ to conclude that $f_z(0) \geq \delta$, proving that f is not constant.

Observe next that the locally uniform convergence of $\{f_{n_j}\}$ to f implies that the corresponding dilatations ω_{n_j} converge to an analytic function ω with $|\omega(z)| \leq 1$ in D, and $\overline{f_{\bar{z}}} = \omega f_z$. But $|\omega(0)| < 1$ by hypothesis, so the maximum modulus theorem ensures that $|\omega(z)| < 1$ in $\mathbb{D}$. This means that f has positive Jacobian and is therefore locally univalent. Because f is the uniform limit of univalent harmonic functions, it then follows from the argument principle for harmonic functions (Section 1.3) that f is univalent in $\mathbb{D}$. Since $f_n(\mathbb{D}) \subset \Omega_n \subset \Omega$, it is clear that $f(\mathbb{D}) \subset \Omega$.

It remains to show that the radial limits $\hat{f}(e^{i\theta})$ lie on Γ almost everywhere. Here we will use the hypothesis that f_n extends to a homeomorphism of $\overline{\mathbb{D}}$ onto $\overline{\Omega_n}$. Let h_n denote the restriction of $\varphi^{-1} \circ f_n$ to the unit circle $\mathbb{T}$. Then h_n is a sense-preserving homeomorphism of $\mathbb{T}$ onto the circle $|z| = r_n$, and its argument $\alpha_n(\theta) = \arg\{h_n(e^{i\theta})\}$ is an increasing function with

$\alpha_n(2\pi) = \alpha_n(0) + 2\pi$. It may be assumed that $0 \le \alpha_n(\theta) < 2\pi$. By the Helly selection theorem (see, for instance, Duren [1], p. 3), some subsequence of $\{\alpha_n(\theta)\}$ converges almost everywhere to a nondecreasing function $\alpha(\theta)$ with $\alpha(2\pi) = \alpha(0) + 2\pi$. Since r_n tends to 1, the corresponding subsequence of $\{h_n(e^{i\theta})\}$ will converge to $h(\theta) = e^{i\alpha(\theta)}$. After an obvious readjustment, we can suppose that $\{h_{n_j}\}$ converges to h. Then it follows that $\{f_{n_j}(e^{i\theta})\}$ converges almost everywhere to $\varphi \circ h(e^{i\theta})$, whose values lie on Γ.

On the other hand, $f_{n_j}(z)$ tends to $f(z)$ uniformly on compact subsets of $\mathbb{D}$. As a bounded harmonic function, $f(re^{i\theta})$ has radial limits $\hat{f}(e^{i\theta})$ almost everywhere. To complete the proof, we will show that

$$\hat{f}(e^{i\theta}) = \varphi \circ h(e^{i\theta}) \qquad \text{a. e.}$$

For this purpose, recall that a harmonic function can be recovered from its boundary values by integration against the Poisson kernel

$$P(z, t) = \operatorname{Re}\left\{ \frac{e^{it} + z}{e^{it} - z} \right\}.$$

Thus,

$$f_{n_j}(z) = \frac{1}{2\pi} \int_0^{2\pi} f_{n_j}(e^{it}) P(z, t)\, dt.$$

Letting j tend to ∞, we conclude by appeal to the Lebesgue bounded convergence theorem that

$$f(z) = \frac{1}{2\pi} \int_0^{2\pi} \varphi \circ h(e^{it}) P(z, t)\, dt.$$

Now take radial limits to arrive at the desired conclusion. ∎

7.6. Proof of Existence

We turn now to the proof of Theorem 1. This is a relatively deep result whose full proof lies beyond the scope of this book. The main ideas were introduced by Bers [2] and Bojarski [5], in the more general context of elliptic partial differential equations, and were later refined by Bojarski and Iwaniec [1]. Broad expositions can be found in the books by Renelt [1] and Wendland [1]. Hengartner and Schober [5] gave an outline of the argument specifically adapted to harmonic mappings and the proof of Theorem 1. We propose to amplify their exposition by giving further details and discussing some general principles that underlie the proof.

Let us begin by recalling the *Cauchy–Green theorem*

$$\int_\Gamma f(z)\,dz = 2i \iint_\Omega \frac{\partial f}{\partial \bar{z}}\,dx\,dy\,, \qquad f \in C^1(\overline{\Omega})\,,$$

where Ω is a Jordan domain with smooth boundary Γ. From this it is an easy step to the *Cauchy–Green integral formula*

$$f(z) = \frac{1}{2\pi i} \int_\Gamma \frac{f(\zeta)}{\zeta - z}\,d\zeta - \frac{1}{\pi} \iint_\Omega \frac{\partial f}{\partial \bar{\zeta}}\frac{1}{\zeta - z}\,d\xi\,d\eta\,, \qquad z \in \Omega\,,$$

where $\zeta = \xi + i\eta$. One interesting application is the formula

$$\bar{z} = -\frac{1}{\pi} \iint_\Omega \frac{1}{\zeta - z}\,d\xi\,d\eta\,, \qquad z \in \Omega\,,$$

which is valid when Ω is a circular disk.

The *Cauchy transform* of a function $f \in L^1(\Omega)$ is defined by

$$(Tf)(z) = -\frac{1}{\pi} \iint_\Omega \frac{f(\zeta)}{\zeta - z}\,d\xi\,d\eta\,, \qquad z \in \Omega\,.$$

The integral can be shown to converge for almost every point z in Ω, and $Tf \in L^p(\Omega)$ for every $p < 2$ (*cf.* Vekua [1]). For $\varphi \in C_0^\infty$, the class of infinitely differentiable functions with compact support in Ω, the Cauchy–Green formula says that $T(\varphi_{\bar{z}}) = \varphi$. A straightforward calculation then shows that the Cauchy transform $F = Tf$ of every function $f \in L^1(\Omega)$ satisfies $F_{\bar{z}} = f$ in the weak sense:

$$\iint_\Omega F \frac{\partial \varphi}{\partial \bar{z}}\,dx\,dy = - \iint_\Omega f\varphi\,dx\,dy\,, \qquad \varphi \in C_0^\infty\,.$$

If f satisfies a Lipschitz condition in $\overline{\Omega}$, a direct calculation shows that $F = Tf$ is differentiable in the ordinary sense, with

$$\frac{\partial F}{\partial \bar{z}} = f \quad \text{and} \quad \frac{\partial F}{\partial z} = -\frac{1}{\pi} \iint_\Omega \frac{f(\zeta)}{(\zeta - z)^2}\,d\xi\,d\eta\,,$$

where the last integral is taken as a Cauchy principal value. For an elegant proof that $\partial F/\partial \bar{z} = f$, one can show that the function $G(z) = F(z) - \bar{z}f(z_0)$ is analytic for each point $z_0 \in \Omega$ by introducing the aforementioned integral formula for $\bar{z}$ and making a direct calculation of the derivative. (Details can be found in Lehto and Virtanen [1], p. 155.)

We turn now to the proof of Theorem 1. Given a function ω analytic in $\mathbb{D}$ and satisfying $|\omega(z)| \leq k < 1$, we are to construct a harmonic mapping f of $\mathbb{D}$ onto Ω with the properties $\overline{f_{\bar{z}}} = \omega f_z$, $f(0) = w_0$, and $f_z(0) > 0$. Here Ω is a Jordan domain with analytic boundary, and w_0 is a prescibed point of Ω.

Suppose first that $\omega(0) = 0$. Let Φ be the conformal mapping of $\mathbb{D}$ onto Ω normalized by $\Phi(0) = w_0$ and $\Phi'(0) > 0$. Since the boundary curve Γ is analytic, Φ has a conformal extension to a larger disk. Make the substitution $f = \Phi \circ g$ and note that

$$f_z = (\Phi' \circ g)g_z, \qquad f_{\bar{z}} = (\Phi' \circ g)g_{\bar{z}}.$$

The existence problem for f is then equivalent to finding a function g that maps $\mathbb{D}$ univalently onto itself, with the properties $g(0) = 0$, $g_z(0) > 0$, and

$$g_{\bar{z}} = \frac{\overline{\Phi' \circ g}}{\Phi' \circ g} \, \omega g_z.$$

Since Φ is univalent, it is clear that $(\Phi' \circ g)(z) \neq 0$ in $\mathbb{D}$.

Now consider the integral operator $\mathcal{P}$ defined by

$$(\mathcal{P}\varphi)(z) = -\frac{1}{\pi} \iint_{\mathbb{D}} \left\{ \frac{\varphi(\zeta)}{\zeta - z} + \frac{z\overline{\varphi(\zeta)}}{1 - \bar{\zeta}z} - \frac{\varphi(\zeta)}{2\zeta} + \frac{\overline{\varphi(\zeta)}}{2\bar{\zeta}} \right\} d\xi \, d\eta,$$
$$\zeta = \xi + i\eta.$$

It can be shown that $\mathcal{P}$ is a compact operator on $L^p(\Omega)$ for $p > 2$. Observe that it has the properties

$$\operatorname{Re}\{(\mathcal{P}\varphi)(e^{i\theta})\} = 0 \qquad \text{and} \qquad \operatorname{Im}\{(\mathcal{P}\varphi)(0)\} = 0.$$

The second term in the integrand contributes a function analytic in $\mathbb{D}$, and the last two terms contribute a constant. The first term gives a Cauchy transform, so $(\mathcal{P}\varphi)_{\bar{z}} = \varphi$ in the weak sense, while

$$(\mathcal{P}\varphi)_z = -\frac{1}{\pi} \iint_{\mathbb{D}} \frac{\varphi(\zeta)}{(\zeta - z)^2} \, d\xi \, d\eta - \frac{1}{\pi} \iint_{\mathbb{D}} \frac{\overline{\varphi(\zeta)}}{(1 - \bar{\zeta}z)^2} \, d\xi \, d\eta.$$

In this formula, the first integral is a Hilbert transform taken as a principal value. For notational convenience, we write $(\mathcal{P}\varphi)_z = \mathcal{H}\varphi$.

Now make the further substitution $g(z) = e^{(\mathcal{P}\varphi)(z)}$. Then $g(0) = 0$ and $g_z(0) = e^{(\mathcal{P}\varphi)(0)} > 0$, since $(\mathcal{P}\varphi)(0)$ is real. Calculations give

$$g_z = e^{\mathcal{P}\varphi} + ze^{\mathcal{P}\varphi}(\mathcal{P}\varphi)_z = e^{\mathcal{P}\varphi}[1 + z(\mathcal{H}\varphi)],$$
$$g_{\bar{z}} = ze^{\mathcal{P}\varphi}(\mathcal{P}\varphi)_{\bar{z}} = z\varphi e^{\mathcal{P}\varphi},$$

so the differential equation for g transforms to the fixed-point problem

$$\varphi = \frac{\overline{\omega}}{z} (\mathcal{M}\varphi) [1 + \overline{z(\mathcal{H}\varphi)}], \qquad (\star)$$

where

$$\mathcal{M}\varphi = \frac{\overline{\Phi'(z \exp\{\mathcal{P}\varphi\})}}{\Phi'(z \exp\{\mathcal{P}\varphi\})} \exp\{-2i \operatorname{Im}\{\mathcal{P}\varphi\}\} \,.$$

Note that $|(\mathcal{M}\varphi)(z)| \equiv 1$. Now since $|\omega(z)| \leq k < 1$ and $\omega(0) = 0$, an argument based on the contraction mapping principle (see Wendland [1], p. 56) shows that, for some $p > 2$, the nonlinear equation

$$\psi = \frac{\overline{\omega}}{z}(\mathcal{M}\varphi)[1 + \overline{z(H\psi)}]$$

has a unique solution $\psi \in L^p(\mathbb{D})$ for each given function $\varphi \in L^p(\mathbb{D})$. The mapping $\varphi \mapsto \psi$ so defined can be shown to be compact and to map some closed ball of $L^p(\mathbb{D})$ into itself. Hence, the Schauder fixed-point theorem (*cf.* Schwartz [1], p. 96) guarantees the existence of a function $\varphi \in L^p(\mathbb{D})$ that satisfies $(\star)$. The function $f = \Phi(ze^{\mathcal{P}\varphi})$ is then a solution of the Beltrami equation $\overline{f_{\overline{z}}} = \omega f_z$. Because $|\omega(z)| \leq k < 1$ and ω is analytic, it follows from the basic theory of quasiconformal mappings that in fact f is infinitely differentiable in the usual sense, and so is a harmonic function.

Since $\operatorname{Re}\{(\mathcal{P}\varphi)(0)\} = 0$ for $z \in \mathbb{T}$, it is clear that $f(\mathbb{T}) \subset \Gamma$. Furthermore, $f(z) - w_0$ vanishes only at the origin, where it has a zero of order 1. An appeal to the argument principle for harmonic functions (*cf.* Section 1.3) therefore shows that f maps $\mathbb{D}$ univalently onto Ω.

If $\omega(0) \neq 0$, we apply what we have already proved to the domain

$$\widetilde{\Omega} = \{w - \overline{\omega(0)}\,\overline{w} \ : \ w \in \Omega\} \,,$$

the image of Ω under an affine mapping, prescribing the dilatation

$$\widetilde{\omega}(z) = \frac{\omega(z) - \omega(0)}{1 - \overline{\omega(0)}\,\omega(z)} \,,$$

for which $\widetilde{\omega}(0) = 0$. We then obtain a harmonic mapping $\widetilde{f} : \mathbb{D} \to \widetilde{\Omega}$ with the properties $\overline{\widetilde{f}_{\overline{z}}} = \widetilde{\omega}\,\widetilde{f}_z$, $\widetilde{f}(0) = \widetilde{w}_0 = w_0 - \overline{\omega(0)}\,\overline{w}_0$, and $\widetilde{f}_z > 0$. Inverting the affine mapping, we see that

$$f = \frac{\widetilde{f} + \overline{\omega(0)}\,\overline{\widetilde{f}}}{1 - |\omega(0)|^2}$$

is a harmonic mapping of $\mathbb{D}$ onto Ω with the properties $f(0) = w_0$ and

$$f_z(0) = \frac{1}{1 - |\omega(0)|^2}\{\widetilde{f}_z(0) + \overline{\omega(0)}\,\overline{\widetilde{f}_{\overline{z}}(0)}\}$$

$$= \frac{1}{1 - |\omega(0)|^2}\,\widetilde{f}_z(0) > 0 \,.$$

Further calculations give

$$(1 - |\omega(0)|^2)f_z = \tilde{f}_z + \overline{\omega(0)}\,\overline{\tilde{f}_{\bar{z}}} = (1 + \overline{\omega(0)}\,\tilde{\omega})\tilde{f}_z$$

$$(1 - |\omega(0)|^2)f_{\bar{z}} = \tilde{f}_{\bar{z}} + \overline{\omega(0)}\,\overline{\tilde{f}_z} = (\tilde{\omega} + \overline{\omega(0)})\overline{\tilde{f}_z}\,,$$

so that

$$\frac{\overline{f_{\bar{z}}}}{f_z} = \frac{\tilde{\omega} + \omega(0)}{1 + \overline{\omega(0)}\,\tilde{\omega}} = \omega\,.$$

Thus, f is the desired mapping with dilatation ω.

7.7. Uniqueness Problem

Concerning the general question of uniqueness of a harmonic mapping onto a specified domain with prescribed analytic dilatation ω, very little is known. This is rather surprising, because the corresponding problem for prescribed *first* complex dilatation $\mu = f_{\bar{z}}/f_z$ is relatively simple and was solved long ago. In fact, a direct calculation shows that a composite mapping $g \circ f$ has the same dilatation μ if and only if g is conformal on the range of f. Thus, a homeomorphism f of $\mathbb{D}$ onto Ω with prescribed dilatation μ is uniquely determined up to postcomposition with a conformal self-mapping of Ω (see Lehto and Virtanen [1], Ch. V, for further details).

The uniqueness problem for prescribed *second* complex dilatation $\omega = \overline{f_{\bar{z}}}/f_z$ is essentially different because ω transforms in a more complicated way under postcomposition. Under *pre*composition with a conformal mapping φ, the analytic dilatation of $f \circ \varphi$ becomes simply $\omega \circ \varphi$, but this is a different function unless ω is constant.

For one class of mappings, however, the uniqueness question has an affirmative answer. A normalized harmonic mapping with prescribed dilatation ω is known to be unique when the target region Ω is *strictly starlike*. This means that Ω contains the origin and that each radial half-line from the origin intersects the boundary of Ω in exactly one point. An equivalent requirement is that $\overline{\Omega} \subset \lambda\Omega$ for each magnification factor $\lambda > 1$. The following theorem appears in a paper by D. Bshouty, N. Hengartner, and W. Hengartner [1] with attribution to Reiner Kühnau. It can be regarded as a subordination principle for starlike harmonic mappings.

Theorem. *Let Ω be a bounded strictly starlike domain and let ω be an analytic function with $|\omega(z)| < 1$ in $\mathbb{D}$. Let f be a harmonic mapping of $\mathbb{D}$ onto Ω with dilatation $\overline{f_{\bar{z}}}/f_z = \omega$ and the normalization $f(0) = 0$, $f_z(0) > 0$. Let g*

be a harmonic function in $\mathbb{D}$ *with the properties* $\overline{g_{\bar{z}}} = \omega g_z$, $g(0) = 0$, *and* $g_z(0) \geq 0$, *whose range* $g(\mathbb{D})$ *is contained in* Ω. *Then* $g_z(0) < f_z(0)$ *unless* $g = f$.

Corollary. *Let* Ω *be a bounded strictly starlike domain and let* ω *be an analytic function with* $|\omega(z)| < 1$ *in* $\mathbb{D}$. *Then there is at most one harmonic mapping* f *of* $\mathbb{D}$ *onto* Ω *with dilatation* $\omega = \overline{f_{\bar{z}}}/f_z$ *that is normalized by the conditions* $f(0) = 0$, $f_z(0) > 0$.

If ω has the stronger property $|\omega(z)| \leq k < 1$, the corollary can be combined with the main existence theorem (Theorem 1 in Section 7.5) to give the existence and uniqueness of a normalized harmonic mapping of $\mathbb{D}$ onto Ω with dilatation ω. It may also be remarked that with the help of the Riemann mapping theorem, the above subordination principle can be transplanted to replace the unit disk by any proper simply connected domain D in the plane. The common normalization then takes the form $f(z_0) = 0$, $f_z(z_0) > 0$ for some arbitrarily chosen point z_0 in D, and the conclusion becomes $g_z(z_0) < f_z(z_0)$.

Proof of Theorem. Fix a positive number $\rho < 1$ and form the function

$$F(z) = g_z(0)f(\rho z) - f_z(0)g(\rho z).$$

Clearly, F is harmonic in $\overline{\mathbb{D}}$ and it has the properties $F(0) = 0$ and $F_z(0) = 0$. Since f and g have the same dilatation, it is also clear that $F_{\bar{z}}(0) = 0$. An easy calculation shows that F satisfies a Beltrami equation of the second kind with dilatation $\omega(\rho z)$; in other words,

$$\overline{F_{\bar{z}}(z)} = \omega(\rho z)F_z(z), \qquad z \in \mathbb{D}.$$

Thus, F is a sense-preserving harmonic function with a zero of order $n \geq 2$ at the origin (see Section 1.3).

Now write $\lambda = g_z(0)/f_z(0)$ and suppose for the purpose of contradiction that $\lambda > 1$. Then, in view of the hypotheses that $g(\mathbb{D}) \subset f(\mathbb{D}) = \Omega$ and Ω is strictly starlike, it is clear geometrically that $\overline{g(\mathbb{D})} \subset \lambda f(\mathbb{D})$. It follows by continuity that $\overline{g(\rho\mathbb{D})} \subset \lambda f(\rho\mathbb{D})$ if ρ is chosen sufficiently close to one. This implies in particular that $F(z) \neq 0$ on the unit circle $\mathbb{T}$, since f is univalent in $\mathbb{D}$ and so the sets $g(\rho\mathbb{T})$ and $\lambda f(\rho\mathbb{T})$ are disjoint. The inclusion also shows that $\arg\{F(z)\}$ increases by exactly 2π as z moves once around $\mathbb{T}$ in the positive (counterclockwise) direction. In self-explanatory notation, $\Delta_{\mathbb{T}} \arg\{F(z)\} = 2\pi$. But this is a violation of the argument principle, since F is a sense-preserving harmonic function with at least double zero at the origin. (For the harmonic version of the argument principle, see Section 1.3.) The contradiction shows that $\lambda \leq 1$, or $g_z(0) \leq f_z(0)$.

It remains to discuss the case of equality. Suppose that $g_z(0) = f_z(0)$ but $g \neq f$. For fixed $t > 1$, consider the function $G = tf - g$. Observe that G is harmonic in $\mathbb{D}$ with the properties $G(0) = 0$, $G_z(0) > 0$, and $\overline{G_{\bar{z}}} = \omega G_z$. Because $t > 1$ and Ω is strictly starlike, the same geometric considerations as before now show that $\Delta_C \arg\{G(z)\} = 2\pi$ for every circle C of the form $|z| = \rho$ with ρ sufficiently close to one. But G has a simple zero at the origin, so it follows from the argument principle that $G(z) \neq 0$ elsewhere in $\mathbb{D}$. Consequently, $\Delta_C \arg\{G(z)\} = 2\pi$ for *every* circle C given by $|z| = \rho$ with $0 < \rho < 1$.

On the other hand, the function G tends uniformly to $G^\star = f - g$ as t decreases to one. Under the supposition that $g_z(0) = f_z(0)$ but $g \neq f$, the function $G^\star$ is a nonconstant harmonic function with dilatation ω and with a zero of order $n \geq 2$ at the origin. Hence, $\Delta_C \arg\{G^\star(z)\} = 2n\pi > 2\pi$ for every small circle C centered at the origin. But this is impossible, because the uniform convergence implies that

$$\Delta_C \arg\{G^\star(z)\} = \lim_{t \to 1} \Delta_C \arg\{G(z)\} = 2\pi.$$

The contradiction shows that $g = f$ if $g_z(0) = f_z(0)$, and the proof is complete. ∎

8

Additional Topics

8.1. Harmonic Mappings of Annuli

It is a classical result of conformal mapping theory that any nondegenerate doubly connected domain can be mapped conformally onto an annulus $\rho_1 < |w| < \rho_2$, and the ratio $M = \rho_2/\rho_1$ is a conformal invariant known as the *modulus* of the given domain. In particular, if an annulus $r_1 < |z| < r_2$ is mapped conformally onto $\rho_1 < |w| < \rho_2$, then $\rho_2/\rho_1 = r_2/r_1$.

The last statement is easily proved. Let $w = f(z)$ be a conformal mapping of the first annulus onto the second. After an inversion (if necessary), which does not change the modulus of the annulus, it may be supposed that f carries the inner boundary circle $|z| = r_1$ onto the inner boundary $w| = \rho_1$. Then by successive Schwarz reflections over the inner and outer boundaries, the function f can be extended to a conformal mapping of $\mathbb{C}\backslash\{0\}$ onto itself. But the extended mapping is bounded near the origin, so the apparent singularity there can be removed by defining $f(0) = 0$. The resulting function f maps the entire complex plane conformally onto itself, so it has the form $f(z) = \alpha z$ for some constant α (see Section 2.4). For another proof based on the maximum principle for harmonic functions, see Nehari [2], p. 333.

The question now arises whether the modulus is still preserved, at least up to constant factors, under harmonic mapping. The answer is surprising. Johannes Nitsche [5] found that a nondegenerate annulus may be mapped harmonically onto a punctured disk, but not onto too "thin" an annulus. In other words, there is a constant $C = C(m) > 1$ depending only on the modulus $m = r_2/r_1$ of the annulus $r_1 < |z| < r_2$, such that $M \geq C$ for any harmonic mapping of it onto an annulus $\rho_1 < |w| < \rho_2$, where $M = \rho_2/\rho_1$ is the modulus of the target annulus.

Let us begin with Nitsche's construction of a harmonic mapping onto a punctured disk. Recall first that in polar coordinates the Laplacian takes the form

$$\Delta u = \frac{\partial^2 u}{\partial r^2} + \frac{1}{r}\frac{\partial u}{\partial r} + \frac{1}{r^2}\frac{\partial^2 u}{\partial \theta^2}.$$

The idea is to look for a harmonic mapping of the special form $f(re^{i\theta}) = \rho(r)e^{i\theta}$, sending a fixed annulus $1 < |z| < R$ onto some annulus $\rho_1 < |w| < \rho_2$. If $\Delta f = 0$, then $\rho(r)$ satisfies the differential equation

$$r^2\rho''(r) + r\rho'(r) - \rho(r) = 0.$$

This is an Euler equation, and the standard method is to try solutions of the form $\rho(r) = r^n$. A short calculation shows that $n = \pm 1$, so the general solution is $\rho(r) = ar + b/r$, where a and b are constants.

The constants must be chosen so that $\rho(r) > 0$ in the interval $1 < r < R$; thus, $a + b \geq 0$. If f is to be univalent, the function $\rho(r)$ must be monotonic; there is no loss in taking it to be strictly increasing. It is therefore required that

$$\rho'(r) = a - b/r^2 > 0, \qquad 1 < r < R;$$

thus, $a - b \geq 0$. It is now seen that $\rho(r) > 0$ and $\rho'(r) > 0$ for $1 < r < R$ if and only if $a > 0$ and $-a \leq b \leq a$.

Under the harmonic mapping

$$w = f(re^{i\theta}) = (ar + b/r)e^{i\theta}$$

with $a > 0$ and $-a \leq b \leq a$, the annulus $1 < |z| < R$ is carried onto $\rho_1 < |w| < \rho_2$, where

$$\rho_1 = a + b, \qquad \rho_2 = aR + b/R.$$

If $b = -a$, then $\rho_1 = 0$ and the range is a punctured disk. At the other extreme, the modulus $M = \rho_2/\rho_1$ attains its minimum value $(1 + R^2)/2R$ when $b = a$. In other words, under harmonic mappings *of this particular type*, an annulus of modulus R is always transformed to an annulus of modulus $M \geq (1 + R^2)/2R$.

The last principle actually extends to arbitrary harmonic mappings between annuli, with the modulus of the range satisfying $M \geq C(R)$ for some constant $C(R) > 1$ depending only on R. To see this, let $f = u + iv$ be any harmonic mapping of an annulus A defined by $1 < |z| < R$ onto an annulus $1 < |w| < M$. Here R is specified but M is not. Let B be a fixed closed subannulus of A; to be specific, let B be defined by $(2 + R)/3 \leq |z| \leq (1 + 2R)/3$. Then it is obvious geometrically that there are points z_1 and z_2 in B where $u(z_1) < -1$ and $u(z_2) > 1$. But $U(z) = u(z) + M$ is a positive harmonic function in A, so Harnack's inequality (see Section 1.4) gives $U(z_2) \leq kU(z_1)$ for some constant $k = k(R) > 1$ depending only on the modulus R of A. Therefore,

$$M + 1 < M + u(z_2) = U(z_2) \leq kU(z_1) = k(M + u(z_1)) < k(M - 1),$$

and so $k > (M + 1)/(M - 1)$. But this gives $M > (k + 1)/(k - 1)$, so that $M \geq C(R) > 1$, as claimed.

It is an interesting open question whether the inequality $M \geq (1 + R^2)/2R$ remains valid for arbitrary harmonic mappings between annuli. A more ambitious problem is to find the sharp value of the lower bound $C(R)$. Lyzzaik [6] obtained a *quantitative* lower bound, showing that $M \geq 1/s$, where s is the length of the slit in the Grötzsch domain (*cf.* Section 5.2) conformally equivalent to the annulus $1 \leq |z| \leq R$. Recall that a Grötzsch domain consists of the unit disk with a segment $0 \leq z \leq s$ removed, where $0 < s < 1$. There was reason to suspect that the lower bound $1/s$ might be sharp, but Weitsman [5] disproved this by showing that $M \geq 1 + \frac{1}{2}(\frac{\log R}{R})^2$, an improvement on Lyzzaik's estimate when R is near 1. It seems unlikely that the inequality $M \geq (1 + R^2)/2R$ extends to the full class of harmonic mappings, but no counterexample has been found.

Bshouty and Hengartner [2,3] have constructed a large family of harmonic mappings, in addition to Nitsche's example, of an annulus onto a punctured disk. Duren and Hengartner [2] have shown more generally that any finitely connected domain can be mapped harmonically onto a prescribed convex region with suitable punctures. In terms of canonical conformal mappings, they have also constructed a harmonic mapping of an arbitrary domain onto a punctured plane. These results will be described in the next section.

8.2. Multiply Connected Domains

The Radó–Kneser–Choquet theorem was discussed in Chapter 3. Its basic assertion is that the unit disk can be mapped harmonically onto any bounded convex domain with prescribed homeomorphic boundary values. In view of the Riemann mapping theorem, the result extends to harmonic mappings of any simply connected Jordan domain onto a bounded convex domain. The assumption of convexity can be relaxed, as Kneser remarked; the essential requirement is that the range of the harmonic extension be contained in the target domain.

Our object is now to extend the Radó–Kneser–Choquet theorem to multiply connected domains. This will be done in two distinct ways. The first generalization is comparatively straightforward. The following theorem is a restricted version of a result of Duren and Hengartner [2], obtained also by Lyzzaik [4].

Theorem 1. *Let D and Ω be bounded finitely connected Jordan domains of equal connectivity. Let φ be a sense-preserving homeomorphism of ∂D onto $\partial \Omega$, and let f be its harmonic extension to D. If $f(D) \subset \Omega$, then f maps D univalently onto Ω. Conversely, if f is univalent in D, then $f(D) = \Omega$.*

The strategy of proof is that of Kneser (see Section 3.1), but the additional boundary components present a new complication. An argument must now be devised to show that the level-set of the function ψ emanating from a critical point cannot have closed loops in D surrounding boundary components. The details will not be pursued here. Instead we shall focus on a second more striking generalization of the Radó–Kneser–Choquet theorem, where convexity plays an essential role. This result is also in the paper by Duren and Hengartner [2].

Theorem 2. *Let D be a simply connected domain bounded by Jordan curves $C_0, C_1, \ldots, C_n$, where C_0 is the outer boundary component. Let Ω be a bounded convex domain, and let φ be a sense-preserving homeomorphism of C_0 onto $\partial\Omega$. Then there is a function f harmonic in D and continuous in $\overline{D}$, with $f(z) = \varphi(z)$ on C_0, mapping D univalently onto Ω with n points removed.*

It must be emphasized that the locations of the punctures are not prescribed independently but will depend on the given boundary function φ. Bshouty and Hengartner [2, 3] proved the theorem for doubly connected domains ($n = 1$), where D may be taken to be an annulus, and they found that the inner boundary component may always be mapped to the average value of φ. This special result will follow from the construction used below to establish the more general theorem.

Proof of Theorem 2. There is no loss of generality in assuming that the boundary curves C_k are analytic, since this can be achieved by a suitable conformal mapping. Let F be the function harmonic in D and continuous in $\overline{D}$, with $F(z) = \varphi(z)$ on C_0 and $F(z) \equiv 0$ on the rest of the boundary. Let ω_j be the harmonic measure of C_j with respect to D, and let

$$p_{jk} = \int_{C_k} \frac{\partial \omega_j}{\partial n}\, ds, \qquad j, k = 0, 1, \ldots, n,$$

denote the period of the harmonic conjugate of ω_j around C_k, where $\partial/\partial n$ denotes the inner normal derivative. Finally, let

$$\gamma_k = \int_{C_k} \frac{\partial F}{\partial n}\, ds, \qquad k = 1, 2, \ldots, n.$$

The periods p_{jk} are conformal invariants that form a symmetric matrix known as the *Riemann matrix* of the domain D. It is well known (see, for instance, Nehari [2], Ch. 1) that the submatrix (p_{jk}) for $j, k = 1, 2, \ldots, n$ is nonsingular. Thus, the linear system of equations

$$\sum_{j=1}^{n} \lambda_j p_{jk} + \gamma_k = 0, \qquad k = 1, 2, \ldots, n,$$

has a unique complex solution (λ_j). We now define

$$f = F + \sum_{j=1}^{n} \lambda_j \omega_j.$$

Then f is harmonic in D and it has boundary values $f(z) = \varphi(z)$ on C_0 and $f(z) \equiv \lambda_k$ on C_k for $k = 1, 2, \ldots, n$. By construction,

$$\int_{C_k} \frac{\partial f}{\partial n} \, ds = 0, \qquad k = 0, 1, \ldots, n.$$

Our aim is to prove that f maps D univalently onto Ω with the n points λ_k removed. The first step is to show that all of the points λ_k actually lie in Ω. Indeed, the maximum principle ensures that, for each angle α, the harmonic function $\mathrm{Re}\,\{e^{i\alpha} f(z)\}$ attains its maximum value on the boundary. If some point λ_k lies outside the *convex* domain Ω, then α can be chosen so that

$$\max_{z \in \overline{D}} \mathrm{Re}\,\{e^{i\alpha} f(z)\} = \mathrm{Re}\,\{e^{i\alpha}\lambda_j\}$$

for some λ_j, since $f(C_0) = \partial\Omega$ and $f(C_k) = \lambda_k$ for $k = 1, 2, \ldots, n$. Preceding f by a suitable conformal mapping, we may take C_j to be a circle $|z - \zeta| = \rho$ with center ζ and radius ρ. By conformal invariance of the flux integrals,

$$r \frac{\partial}{\partial r} \int_0^{2\pi} f(\zeta + re^{i\theta}) \, d\theta = \int_{C_j} \frac{\partial f}{\partial n} \, ds = 0.$$

Since $f(z) \equiv \lambda_j$ on C_j, this implies that

$$\frac{1}{2\pi} \int_0^{2\pi} f(\zeta + re^{i\theta}) \, d\theta \equiv \lambda_j, \qquad \rho \leq r < \rho + \varepsilon,$$

for some $\varepsilon > 0$. But since $\mathrm{Re}\,\{e^{i\alpha} f(z)\} \leq \mathrm{Re}\,\{e^{i\alpha}\lambda_j\}$ in D, it then follows that $\mathrm{Re}\,\{e^{i\alpha} f(z)\} \equiv \mathrm{Re}\,\{e^{i\alpha}\lambda_j\}$ in the annulus $\rho < |z| < \rho + \varepsilon$, hence by harmonic continuation throughout the domain D. But this is a contradiction, because f maps C_0 onto $\partial\Omega$ and so its values in D do not lie on a line. Thus we conclude that all of the points λ_k are in the domain Ω.

The next step is to show that f has a *global* representation $f = h + \overline{g}$, where h and g are analytic in D. This is a consequence of the property

$$\int_{C_k} \frac{\partial f}{\partial n} \, ds = 0, \qquad k = 0, 1, \ldots, n,$$

which says that f has a single-valued conjugate function in D. But if $f = h + \overline{g}$ is some local representation, then it follows from the identities $\mathrm{Re}\,\{f\} = \mathrm{Re}\,\{h + g\}$ and $\mathrm{Im}\,\{f\} = \mathrm{Im}\,\{h - g\}$ that $h + g$ and $h - g$ have single-valued analytic extensions to the whole domain D. Thus, h and g extend to analytic functions in D, and f has a global representation $f = h + \overline{g}$.

It remains to show that f is univalent in D. For this purpose we define the analytic functions

$$\phi_\alpha = e^{i\alpha} h - e^{-i\alpha} g = e^{i\alpha} f - 2\mathrm{Re}\,\{e^{-i\alpha} g\},$$

where α is a real parameter. Since the boundary components C_j are analytic curves and f is constant on C_j for $1 \le j \le n$, we see that f admits a harmonic continuation and so h and g admit analytic continuations across each of these curves C_j. Thus the images $\phi_\alpha(C_j)$ are *bounded* horizontal segments, $j = 1, 2, \ldots, n$. On the other hand, because f maps C_0 onto the boundary of a convex region, it follows that $\mathrm{Im}\,\{\phi_\alpha\} = \mathrm{Im}\,\{e^{i\alpha} f\}$ is nondecreasing on one arc of C_0 and nonincreasing on the complementary arc. Since ϕ_α is an open mapping, this implies that $\phi_\alpha(C_0)$ is a curve in the extended complex plane bounding a domain convex in the horizontal direction, containing all of the closed segments $\phi_\alpha(C_j)$ for $1 \le j \le n$. Let G denote the domain inside $\phi_\alpha(C_0)$ with those segments removed.

To see that ϕ_α is univalent in D and maps it onto G, we apply the argument principle. Fixing an arbitrary point $w_0 \notin \partial G$, we have only to observe that

$$\int_{\partial D} d\arg\{\phi_\alpha(z) - w_0\} = \int_{C_0} d\arg\{\phi_\alpha(z) - w_0\} = \begin{cases} 2\pi, & w_0 \in G \\ 0, & w_0 \notin G. \end{cases}$$

Since ϕ_α is univalent, it follows that

$$\phi_\alpha'(z) = e^{i\alpha} h'(z) - e^{-i\alpha} g'(z) \ne 0$$

in D for every $\alpha \in \mathbb{R}$, so that $|h'(z)| \ne |g'(z)|$. Thus the dilatation $\omega = g'/h'$ of f satisfies $|\omega(z)| \ne 1$ in D. But $f = \varphi$ is a sense-preserving homeomorphism of C_0 onto $\partial\Omega$, so we deduce that $|\omega(z)| < 1$ in D. This allows us to apply the argument principle to f (see Section 1.3). For $w_0 \notin \partial\Omega$ and $w_0 \notin \lambda_j$ for $j = 1, 2, \ldots, n$, it is evident that

$$\int_{\partial D} d\arg\{f(z) - w_0\} = \int_{C_0} d\arg\{\varphi(z) - w_0\} = \begin{cases} 2\pi, & w_0 \in \Omega \\ 0, & w_0 \notin \Omega. \end{cases}$$

Hence, we conclude that f is univalent in D and maps it onto Ω with the points λ_k removed. Since f preserves the connectivity of D, the n points λ_k must all be distinct. This completes the proof of Theorem 2. ∎

In sharp contrast with the behavior of conformal mappings, another theorem by Duren and Hengartner [2] asserts that any domain can be mapped harmonically onto a punctured plane. Given an arbitrary domain $D \subset \hat{\mathbb{C}}$ containing the point at infinity, we call a function f a *canonical harmonic punctured-plane mapping* if it provides a harmonic mapping of D onto a domain of the form

$$\Omega = \hat{\mathbb{C}} \setminus \bigcup_{j \in J} \{\lambda_j\}$$

for some points $\lambda_j \in \mathbb{C}$, and if it has the form $f(z) = z + o(1)$ near infinity; that is,

$$\lim_{z \to \infty} \{f(z) - z\} = 0.$$

In particular, $f(\infty) = \infty$. Observe that the univalence of f and the normalization at infinity imply (by Lewy's theorem) that f is sense-preserving in D.

Theorem 3. *Every domain $D \subset \hat{\mathbb{C}}$ containing the point at infinity admits a canonical harmonic punctured-plane mapping $F = H + \overline{G}$, where H and G are analytic (and single-valued) in D. If the boundary ∂D has countably many components, such a mapping is unique.*

The simplest example is the domain $\Delta = \{z \in \hat{\mathbb{C}} : |z| > 1\}$, the exterior of the closed unit disk. It is easy to see that

$$f(z) = z - \frac{1}{\overline{z}} = \left(r - \frac{1}{r} \right) e^{i\theta}, \qquad z = r e^{i\theta},$$

maps Δ univalently onto $\hat{\mathbb{C}} \setminus \{0\}$, the extended plane punctured at the origin. More generally, Hengartner and Schober [8] found that a function f provides a sense-preserving harmonic mapping of Δ onto a punctured sphere $\hat{\mathbb{C}} \setminus \{\lambda\}$ if and only if it has the form

$$f(z) = \lambda + a[z + bc\overline{z} + 2(b + c) \log |z| - bc/z - 1/\overline{z}]$$

for complex constants a, b, c where $a \neq 0$, $|b| < 1$, and $|c| \leq 1$. If we add the requirement that $f(z) = z + o(1)$ near infinity, then $\lambda = 0$, $a = 1$, and $b = c = 0$. Thus, $f(z) = z - 1/\overline{z}$ is the uniquely determined canonical harmonic punctured-plane mapping of Δ.

Proof of Theorem 3. To simplify the discussion, we shall treat only the special case where D is finitely connected. Then there is no loss of generality in assuming that D is a Jordan domain with analytic boundary.

Let J_α denote the canonical conformal mapping of D onto a parallel slit domain of inclination α with the positive real axis, normalized at infinity by $J_\alpha(z) = z + o(1)$. (For the existence and uniqueness of this mapping, and special properties, see Ahlfors [3], Goluzin [1], or Nehari [2].) In particular, J_0 maps D conformally onto a horizontal slit domain, whereas $J_{\pi/2}$ maps it onto a vertical slit domain. Furthermore, the remarkable identity

$$J_\alpha(z) = e^{i\alpha}[\cos\alpha J_0(z) - i\sin\alpha J_{\pi/2}(z)]$$

holds for $0 \le \alpha \le \pi$.

Now define the function $F(z) = \mathrm{Re}\,\{J_{\pi/2}(z)\} + i\,\mathrm{Im}\,\{J_0(z)\}$. Note that F is harmonic in D, and it maps each component of ∂D to a single point. It has the standard normalization $F(z) = z + o(1)$ near infinity. Finally, F has a global representation $F = H + \overline{G}$, where

$$H(z) = \frac{1}{2}[J_{\pi/2}(z) + J_0(z)] \qquad \text{and} \qquad G(z) = \frac{1}{2}[J_{\pi/2}(z) - J_0(z)]$$

are single-valued analytic functions in D.

To prove the univalence of F, we consider again the function

$$\Phi_\alpha = e^{i\alpha}H - e^{-i\alpha}G = e^{i\alpha}F - 2\mathrm{Re}\,\{e^{-i\alpha}G\},$$

where α is a real parameter. The formulas for H and G show that

$$\Phi_\alpha(z) = \cos\alpha J_0(z) + i\sin\alpha J_{\pi/2}(z) = e^{i\alpha}J_{-\alpha}(z).$$

In particular, Φ_α is univalent and so $\Phi_\alpha'(z) \ne 0$ in D for each $\alpha \in \mathbb{R}$. This implies that $|H'(z)| + |G'(z)| \ne 0$ and $|\omega(z)| \ne 1$, where $\omega = G'/H'$ is the dilatation of F. But the normalizations of J_0 and $J_{\pi/2}$ at infinity show that $H'(\infty) = 1$ and $G'(\infty) = 0$, so $\omega(\infty) = 0$ and it follows that $|\omega(z)| < 1$ in D.

We claim now that the composed function $Q = F \circ J_0^{-1}$ is univalent on the horizontal slit domain $J_0(D)$. To see this, write $w = u + iv = Q(\zeta)$, where $\zeta = \xi + i\eta$. Observe first that $v = \eta$ by the definitions of Q and F. Thus, Q sends each horizontal line into itself. To determine the action of Q on a given horizontal line, we calculate the derivative

$$\frac{\partial u}{\partial \xi} = \mathrm{Re}\Big\{\frac{\partial}{\partial \xi}J_{\pi/2}\left(J_0^{-1}(\zeta)\right)\Big\} = \mathrm{Re}\,\{J_{\pi/2}'(z)/J_0'(z)\},$$

where $z = J_0^{-1}(\zeta)$. But $J_{\pi/2}' = H' + G'$ and $J_0' = H' - G'$, so

$$\frac{\partial u}{\partial \xi} = \mathrm{Re}\Big\{\frac{1 + \omega(z)}{1 - \omega(z)}\Big\} > 0, \qquad z = J_0^{-1}(\zeta),$$

since $|\omega(z)| < 1$ in D. This shows that Q has a univalent restriction to each horizontal line that lies entirely in the horizontal slit domain $J_0(D)$, and that Q is univalent on each horizontal half-line or segment that ends at a boundary point of $J_0(D)$. No two such images can overlap, because F sends each component of ∂D to a point. Thus, Q is univalent in $J_0(D)$, performing a horizontal shear that collapses each boundary component to a single point. It follows that F is univalent in D, so it is a canonical punctured-plane mapping.

To prove the uniqueness, suppose that $F_1 = H_1 + \overline{G_1}$ and $F_2 = H_2 + \overline{G_2}$ are two canonical punctured-plane mappings of D. Because $F_k(\partial D) = \partial F_k(D)$ is a finite set of points $(k = 1, 2)$, the maximum principle shows that the harmonic function $f = F_1 - F_2$ is bounded in D, since $f(\infty) = 0$. Note also that f is constant on each component of ∂D.

Consider now the analytic functions

$$\Phi_{k,\alpha} = e^{i\alpha} H_k - e^{-i\alpha} G_k = e^{i\alpha} F_k - 2 \operatorname{Re} \{e^{-i\alpha} G_k\}, \qquad k = 1, 2,$$

and let

$$\Psi_\alpha = \Phi_{1,\alpha} - \Phi_{2,\alpha} = e^{i\alpha} h - e^{-i\alpha} g = e^{i\alpha} f - 2 \operatorname{Re} \{e^{-i\alpha} g\},$$

where $f = h + \overline{g}$. The functions $\Phi_{k,\alpha}$ are again univalent in D, and $\Phi_{k,\alpha}(\infty) = \infty$, so the two sets $\Phi_{k,\alpha}(\partial D)$ are bounded for each value of α. Since $\Psi_\alpha(\infty) = 0$, we conclude from the maximum principle that $\Psi_\alpha(z)$ is bounded in D. However, $\operatorname{Im} \{\Psi_\alpha(z)\}$ is constant on each component of ∂D, which is incompatible with the boundedness of the range unless Ψ_α is constant. Thus, $\Psi_\alpha(z) \equiv 0$ in D for every α, which implies that $h(z) \equiv 0$ and $g(z) \equiv 0$ in D. This shows that $F_1 = F_2$, and Theorem 3 is proved. ∎

If D has infinitely many boundary components, the strategy of proof is to represent it as the union of an expanding sequence of finitely connected Jordan domains with analytic boundary. Uniqueness of the canonical punctured-plane mapping cannot be expected in the general case, since the normalized conformal parallel slit mappings need not be unique if the boundary has uncountably many components.

It is shown by Duren and Hengartner [2] that the dilatation $\omega = g'/h'$ of a canonical harmonic punctured-plane mapping sends D onto the $2n$-times covered unit disk, where n is the connectivity of D. It is also proved that no other normalized univalent mapping of D can have the same dilatation.

Hengartner and Schober [3] have proved the existence of normalized harmonic mappings onto parallel slit domains, with dilatation suitably prescribed.

8.3. Inverse of a Harmonic Mapping

When is the inverse of a harmonic mapping also harmonic? Certainly this is true when the given mapping is analytic or affine. Are there other examples?

In his 1945 paper, Choquet [1] recorded the "simple example" $w = f(z)$ defined by

$$u = x, \qquad \tan v \tan y = \tanh x,$$

where $z = x + iy$ and $w = u + iv$. It is easy to verify (for instance) that f maps the half-strip $\{z : x > 0, 0 < y < \frac{\pi}{2}\}$ univalently onto itself. Indeed, each vertical segment $x = x_0$ in this region is mapped onto itself univalently, since v decreases from $\frac{\pi}{2}$ to 0 when y increases from 0 to $\frac{\pi}{2}$. This also shows that f is sense-reversing. Furthermore, it is clear from the symmetry that the mapping is its own inverse: $f(f(z)) \equiv z$. A straightforward but tedious calculation of the Laplacian confirms that $v = \tan^{-1}(\tanh x / \tan y)$ is a harmonic function.

At the end of his paper, Choquet reported that Jacques Deny had shown this to be essentially the only nontrivial example of a harmonic mapping with harmonic inverse. However, no proof appeared in print until 1987, when Edgar Reich [1] studied a more general problem and characterized the class of harmonic mappings f that admit a nonaffine harmonic mapping g such that the composition $g \circ f$ is again harmonic. Reich also recovered the Choquet–Deny result via the following description of the class of harmonic mappings with harmonic inverse.

Theorem. *Let f be a sense-preserving harmonic mapping in some simply connected domain $\Omega \subset \mathbb{C}$, and suppose that f is neither analytic nor affine. Then the inverse mapping is harmonic if and only if f has the form*

$$f(z) = \alpha\{\beta z + 2i \arg(\gamma - e^{\beta z})\} + \delta,$$

where α, β, γ, δ are complex constants with $\alpha\beta\gamma \neq 0$ and $|e^{-\beta z}| < |\gamma|$ for all $z \in \Omega$.

Before embarking on the proof, let us attempt to correlate Reich's theorem with Choquet's example. Introducing the expressions

$$\tanh x = \frac{e^x - e^{-x}}{e^x + e^{-x}}, \qquad \tan y = -i\frac{e^{iy} - e^{-iy}}{e^{iy} + e^{-iy}},$$

$$\tan^{-1}\xi = -\frac{i}{2}\log\left(\frac{1 + i\xi}{1 - i\xi}\right),$$

and making a lengthy calculation, one arrives at the remarkably simple formula

$$v = \tan^{-1}\left(\frac{\tanh x}{\tan y}\right) = \frac{\pi}{2} - y - \arg(1 - e^{-2z}), \qquad 0 < y < \frac{\pi}{2},$$

where $z = x + iy$. Thus, the complex conjugate of Choquet's (sense-reversing) mapping takes the form

$$f(z) = x - iv = z + i \arg(1 - e^{-2z}) - i\frac{\pi}{2}, \qquad 0 < y < \frac{\pi}{2},$$

which agrees with the formula of the above theorem under the choice of parameters $\alpha = \frac{1}{2}, \beta = 2, \gamma = 1, \delta = -i\frac{\pi}{2}$.

It is also of interest to calculate the dilatation of the general function f displayed in the theorem. Writing

$$f(z) = \alpha\{\beta z + \log(\gamma - e^{-\beta z}) - \log(\overline{\gamma} - e^{-\overline{\beta z}})\} + \delta,$$

we find

$$f_z(z) = \frac{\alpha\beta\gamma}{\gamma - e^{-\beta z}}, \qquad \overline{f_{\overline{z}}(z)} = -\frac{\overline{\alpha}\beta e^{-\beta z}}{\gamma - e^{-\beta z}}.$$

The dilatation of f is therefore

$$\omega(z) = \overline{f_{\overline{z}}(z)}/f_z(z) = -\frac{\overline{\alpha}}{\alpha\gamma}e^{-\beta z},$$

and $|\omega(z)| < 1$ in Ω because of the requirement that $|e^{-\beta z}| < |\gamma|$.

Proof of Theorem. Let $f = h + \overline{g}$ be the canonical decomposition of f into a sum of analytic and antianalytic functions. Then the Jacobian

$$J(z) = |h'(z)|^2 - |g'(z)|^2 > 0, \qquad z \in \Omega,$$

since f is sense-preserving. Also $g'(z) \not\equiv 0$ because of the assumption that f is not analytic.

Let $z = \varphi(w)$ denote the inverse mapping, so that $\varphi(f(z)) = z$. Applying the chain rule for the deriviatives $\partial/\partial z$ and $\partial/\partial\overline{z}$, we find

$$\varphi_w h' + \varphi_{\overline{w}}\overline{g'} = 1, \qquad \varphi_w g' + \varphi_{\overline{w}}\overline{h'} = 0,$$

and a bit of manipulation gives

$$\varphi_w(f(z)) = \frac{1}{J(z)}\overline{h'(z)}, \qquad \varphi_{\overline{w}}(f(z)) = -\frac{1}{J(z)}\overline{g'(z)}.$$

Further differentiation of $\varphi_{\overline{w}}(f(z))$ with respect to z and $\overline{z}$ produces the formulas

$$J^2(\varphi_{\overline{w}w}h' + \varphi_{\overline{w}\overline{w}}g') = \overline{h'}h''g' - \overline{g'}^2 g''$$

and

$$J^2(\varphi_{\overline{w}w}\overline{g'} + \varphi_{\overline{w}\overline{w}}\overline{h'}) = h'\overline{h''}\,\overline{g'} - |h'|^2\overline{g''}.$$

Eliminating the $\varphi_{\overline{w}\overline{w}}$ terms, we find

$$J^3\varphi_{\overline{w}w} = \overline{h'}^2 h''g' - \overline{h'}\,\overline{g'}^2 g'' - h'\overline{h''}|g'|^2 + |h'|^2 g'\overline{g''}.$$

Thus, φ is harmonic ($\varphi_{\overline{w}w} = 0$) if and only if

$$h'\overline{h''}|g'|^2 - |h'|^2 g'\overline{g''} = \overline{h'}^2 h''g' - \overline{h'}\,\overline{g'}^2 g''.$$

After division by $|h'g'|^2$, this last equation becomes

$$\frac{\overline{h''}}{\overline{h'}} - \frac{\overline{g''}}{\overline{g'}} = \frac{\overline{h'}h''}{g'h'} - \frac{\overline{g'}g''}{h'g'},$$

which reduces to an equation of the form

$$(\overline{\lambda} + \mu)h' = (\lambda + \overline{\mu})g',$$

where

$$\lambda = \frac{h''}{h'g'}, \qquad \mu = \frac{g''}{h'g'}.$$

But since $|h'(z)| > |g'(z)|$, this implies that $\overline{\lambda} + \mu = 0$, or $\mu(z) = -\overline{\lambda(z)}$ in Ω. However, both λ and μ are meromorphic functions, so it follows that both are constant. Thus,

$$h'' = \lambda h'g', \qquad g'' = -\overline{\lambda}h'g',$$

where λ is a constant. Observe now that $\lambda \neq 0$, since f was assumed not to be affine.

The next step is to eliminate g from the two differential equations, obtaining a single differential equation for h. Logarithmic differentiation leads to the relation

$$-\overline{\lambda}h' = \frac{g''}{g'} = \frac{h'''}{h''} - \frac{h''}{h'},$$

whereupon another integration yields

$$-\overline{\lambda}h = \log h'' - \log h' + c,$$

where c is a constant. Exponentiation gives $ke^{-\bar\lambda h}h' = h''$, where k is a nonzero constant. Integration now produces the formula $h' = a(e^{-\bar\lambda h} + b)$, where a and b are constants and $a \neq 0$. If $b \neq 0$, separation of variables and a final integration gives

$$\log(1 + be^{\bar\lambda h(z)}) = Az + C, \qquad A \neq 0,$$

or

$$\bar\lambda h(z) = \log\left(\frac{1}{b}(Ke^{Az} - 1)\right), \qquad K \neq 0.$$

The function g can now be computed from the differential equation $h'' = \lambda h'g'$, or

$$\lambda g'(z) = \frac{h''(z)}{h'(z)} = -\frac{A}{Ke^{Az} - 1}.$$

Integration gives

$$\lambda g(z) = -\log(K - e^{-Az}) + D,$$

where D is a constant. Combining this with the formula for $h(z)$, we have

$$f(z) = h(z) + \overline{g(z)} = (1/\bar\lambda)\{Az + 2i\arg(K - e^{-Az})\} + \delta,$$

which agrees with the expression given in the theorem if we define $\alpha = 1/\bar\lambda$, $\beta = A$, and $\gamma = K$.

Finally, if $b = 0$, the differential equation for h reduces to $h' = ae^{-\bar\lambda h}$, which is integrated to give

$$\bar\lambda h(z) = \log(Az + C), \qquad A \neq 0.$$

Then

$$\bar\lambda h'(z) = \frac{A}{Az + C}$$

and

$$\lambda g'(z) = \frac{h''(z)}{h'(z)} = -\frac{A}{Az + C}.$$

The harmonic function $f = h + \bar g$ then has dilatation $\omega = g'/h'$ with $|w(z)| \equiv 1$, and so f cannot be univalent. This shows that $b \neq 0$, completing the proof of the theorem. ∎

8.4. Decomposition of Harmonic Functions

If an analytic function is postcomposed with a harmonic function, the resulting function is harmonic. Moreover, the composition of an analytic function with a sense-preserving (univalent) harmonic mapping is a sense-preserving harmonic function. The question arises whether *every* sense-preserving harmonic function admits such a decomposition. The answer is no, but Duren and Hengartner [1] found a simple necessary and sufficient condition for a representation of this type in terms of the dilatation. Recall that a nonconstant harmonic function f is said to be sense-preserving if its dilatation $\omega = \overline{f_{\bar{z}}}/f_z$ is analytic and satisfies $|\omega(z)| < 1$.

Theorem. *Let f be a complex-valued nonconstant harmonic function defined on a domain $D \subset \mathbb{C}$, and let ω be its dilatation. Then, in order for f to have a representation of the form $f = F \circ \varphi$ for some function φ analytic in D and some sense-preserving harmonic mapping F on $\varphi(D)$, it is necessary and sufficient that $|\omega(z)| < 1$ on D and $\omega(z_1) = \omega(z_2)$ wherever $f(z_1) = f(z_2)$. Under these conditions the representation is unique up to conformal mapping; any other representation $f = \tilde{F} \circ \tilde{\varphi}$ has the form $\tilde{F} = F \circ \psi^{-1}$ and $\tilde{\varphi} = \psi \circ \varphi$ for some conformal mapping ψ defined on $\varphi(D)$.*

Before turning to the proof (which will not be given in full detail), it is instructive to consider two simple examples.

Example 1. The harmonic polynomial $f(z) = z^2 + \frac{2}{3}\bar{z}^{-3}$ has dilatation $\omega(z) = z$, so f is sense-preserving in the unit disk $\mathbb{D}$. We claim, however, that f has no decomposition of the specified form in any neighborhood of the origin. Suppose, on the contrary, that $f = F \circ \varphi$ where φ is analytic near the origin and F is a sense-preserving harmonic univalent function on the range of φ. There is no loss of generality in supposing that $\varphi(0) = 0$. Then F has a representation $F = H + \overline{G}$ near the origin, where H and G are analytic and have Taylor series expansions

$$H(\zeta) = \sum_{n=1}^{\infty} A_n \zeta^n, \qquad G(\zeta) = \sum_{n=1}^{\infty} B_n \zeta^n,$$

with $|A_1| > |B_1| \geq 0$. Since $f = F \circ \varphi$, we see that

$$H(\varphi(z)) = z^2, \qquad G(\varphi(z)) = \frac{2}{3}z^3.$$

The first identity shows that φ has an expansion of the form

$$\varphi(z) = c_2 z^2 + c_3 z^3 + \cdots, \qquad c_2 = 1/A_1,$$

which implies that $G \circ \varphi$ has a power-series expansion beginning with an even power of z. But $G(\varphi(z)) = \frac{2}{3}z^3$, so we have arrived at a contradiction.

Note that the Jacobian of f is $J_f(z) = 4|z|^2(1 - |z|^2)$, which vanishes at the origin, so f fails to be univalent in any neighborhood of the origin. On the other hand, the dilatation $\omega(z) = z$ is univalent. Thus, in every neighborhood of the origin there are points z_1 and z_2 where $f(z_1) = f(z_2)$ but $\omega(z_1) \neq \omega(z_2)$, which is in agreement with the theorem.

Example 2. The harmonic polynomial $f(z) = z^2 + \frac{1}{2}\bar{z}^4$ has dilatation $\omega(z) = z^2$, so again f is sense-preserving in $\mathbb{D}$. But here f does have a representation $f = F \circ \varphi$ of the specified form in $\mathbb{D}$, with $F(\zeta) = \zeta + \frac{1}{2}\bar{\zeta}^2$ and $\varphi(z) = z^2$. The function F is univalent in $\mathbb{D}$, so $f(z_1) = f(z_2)$ implies $z_1^2 = z_2^2$, or $\omega(z_1) = \omega(z_2)$.

Proof of Theorem. Suppose first that $f = F \circ \varphi$, where φ is analytic in D and F is a sense-preserving harmonic mapping on $\varphi(D)$. Then the dilatation ν of F is related to the dilatation ω of f by the formula $\omega(z) = \nu(\varphi(z))$, $z \in D$. Because F is univalent, $f(z_1) = f(z_2)$ implies $\varphi(z_1) = \varphi(z_2)$, which implies $\omega(z_1) = \omega(z_2)$.

The converse is more difficult. Here we suppose that f is a nonconstant sense-preserving harmonic function, so that its dilatation satisfies $|\omega(z)| < 1$ in D. We suppose further that $\omega(z_1) = \omega(z_2)$ wherever $f(z_1) = f(z_2)$. For simplicity we will treat only the special case where $|\omega(z)| \leq k < 1$.

We first observe that the problem reduces to finding a univalent function G on $f(D)$ for which the composition $\varphi = G \circ f$ is analytic in D. Then we can define $F = G^{-1}$ to get $f = F \circ \varphi$, where F is univalent on $\varphi(D)$. Near any point ζ where $\varphi'(\zeta) \neq 0$, we can then conclude that $F = f \circ \varphi^{-1}$ is harmonic, where φ^{-1} is a local inverse. But F is locally bounded, so the (isolated) images of critical points of φ are removable singularities, and F is harmonic in $\varphi(D)$.

Thus, it suffices to produce a univalent function G for which $(G \circ f)_{\bar{z}} = 0$, or $G_w f_{\bar{z}} + G_{\bar{w}} \overline{f_z} = 0$, which reduces to the Beltrami equation $G_{\bar{w}} = \mu G_w$, where

$$\mu(w) = -\overline{\omega(f^{-1}(w))}.$$

Although f may not be univalent, and so f^{-1} may be multiple-valued, the hypothesis that $\omega(z_1) = \omega(z_2)$ wherever $f(z_1) = f(z_2)$ ensures that $w \circ f^{-1}$ is single-valued. Thus the function μ is well-defined and it has the property

$|\mu(w)| \le k < 1$ in $f(D)$. We now invoke a standard result from the theory of quasiconformal mappings (see Lehto and Virtanen [1], Ch. 5), which guarantees the existence of a homeomorphism G with prescribed first complex dilatation $\mu = G_{\overline{w}}/G_w$.

To prove the uniqueness assertion, suppose f has another representation $f = \tilde{F} \circ \tilde{\varphi}$ of the same type. Then $\tilde{G} = \tilde{F}^{-1}$ is a smooth function for which $\tilde{\varphi} = \tilde{G} \circ f$ is analytic and nonconstant, so $\tilde{G}$ satisfies the same Beltrami equation: $\tilde{G}_{\overline{w}} = \mu \tilde{G}_w$. But by the uniqueness of quasiconformal mappings with prescribed first complex dilatation (*cf.* Lehto and Virtanen [1], Ch. 4), we may then conclude that $\tilde{G} = \psi \circ G$ for some conformal mapping ψ of $\varphi(D)$. Thus, $\tilde{F} = F \circ \psi^{-1}$ and $\tilde{\varphi} = \psi \circ \varphi$, as the theorem asserts. ∎

In the general case, where it is assumed only that $|\omega(z)| < 1$ in D, the domain D is expressed as an expanding union of compact subsets, in each of which the above construction applies, and the desired representation is obtained by a limiting process. Details are given by Duren and Hengartner [1].

8.5. Integral Means

A function f analytic in the unit disk is said to belong to the *Hardy space* H^P if the integral means

$$M_p(r, f) = \left\{ \frac{1}{2\pi} \int_0^{2\pi} |f(re^{i\theta})|^p d\theta \right\}^{1/p}$$

are bounded for $0 < r < 1$. The space h^p consists of all harmonic functions u for which $M_p(r, u)$ is bounded. Here p is any positive index, $0 < p < \infty$.

If $f \in H^p$ for some $p > 0$, it is known that the radial limits

$$f(e^{i\theta}) = \lim_{r \to 1} f(re^{i\theta})$$

exist for almost every $\theta \in [0, 2\pi]$. If $f = u + iv$ belongs to H^p, then clearly u and v are in h^p. If $u \in h^p$ for some p in the range $1 < p < \infty$, then by a theorem of M. Riesz it follows that $f \in H^p$. This is not true for other values of p. A theorem of Kolmogorov says that if $u \in h^1$, then $f \in H^p$ for all $p < 1$. However, there exist functions $u \in h^p$ for all $p < 1$ such that $f(z)$ has a radial limit in almost no direction, so that $f \notin H^p$ for any $p > 0$.

If f is analytic and *univalent* in $\mathbb{D}$, then it can be shown that $f \in H^p$ for all $p < \frac{1}{2}$. The Koebe function $k(z) = z(1 - z)^{-2}$ shows that the bound $\frac{1}{2}$

is sharp, since $k \notin H^{1/2}$. Proofs of these statements and further information about Hardy spaces and integral means can be found in Duren [1].

As a partial generalization of the result that analytic univalent functions are in H^p for all $p < \frac{1}{2}$, it can be shown that, for harmonic univalent functions $f = h + \overline{g}$ in the disk, the functions h and g are in H^p for *some* positive values of p, so that $f \in h^p$. In particular, each harmonic univalent function in the disk has a radial limit in almost every direction.

For a more precise statement of the result, let $A = \sup |a_2|$ over all functions $f \in S_H$. As shown in Section 6.3, the coefficients a_2 of functions $f \in S_H^0$ satisfy $|a_2| < 48.4$, so $A < 48.4 + \frac{1}{2} < 49$. Then a theorem by Abu-Muhanna and Lyzzaik [2], as sharpened by Nowak [1], is as follows.

Theorem 1. *If $f = h + \overline{g}$ is a harmonic univalent function in the unit disk, then $h \in H^p$, $g \in H^p$, and consequently $f \in h^p$, for all $p < A^{-2}$. Thus $f \in h^p$ for all $p < 0.0004$.*

Abu-Muhanna and Lyzzaik derived the bound $p < (2A + 2)^{-2}$, whereas Nowak improved it to $p < A^{-2}$. The harmonic Koebe function $K = H + \overline{G}$ has its components H and G in the space H^p for all $p < \frac{1}{3}$, but neither H nor G is in $H^{1/3}$. This suggests the possibility that for *all* harmonic univalent functions $f = h + \overline{g}$, it may be true that $h \in H^p$ and $g \in H^p$ for all $p < \frac{1}{3}$. Cima and Livingston [1] verified this for starlike functions, and Nowak [1] extended the result to all close-to-convex harmonic mappings. For convex mappings, Nowak showed the sharp bound to be $\frac{1}{2}$. Thus she established the following theorem.

Theorem 2. *Let $f = h + \overline{g}$ be harmonic and univalent in $\mathbb{D}$. If f is close to convex, then $h \in H^p$, $g \in H^p$, and $f \in h^p$ for all $p < \frac{1}{3}$. If f is convex, then $h \in H^p$, $g \in H^p$, and $f \in h^p$ for all $p < \frac{1}{2}$. The bounds $\frac{1}{3}$ and $\frac{1}{2}$ are sharp.*

Corollary. *If $f = h + \overline{g}$ is a harmonic mapping of $\mathbb{D}$ onto a starlike domain, then $h \in H^p$, $g \in H^p$, and $f \in h^p$ for all $p < \frac{1}{3}$.*

Nowak [1] made the surprising discovery that the harmonic Koebe function K is actually in the space $h^{1/3}$, so it may even be true that all harmonic univalent functions belong to $h^{1/3}$. Likewise she showed that the canonical half-plane mapping $L = h + \overline{g}$ defined (as in Section 3.4) by

$$h(z) = \frac{1}{2}[\ell(z) + k(z)], \qquad g(z) = \frac{1}{2}[\ell(z) - k(z)],$$

belongs to the class $h^{1/2}$, although neither h nor g is in $H^{1/2}$, and $L \notin h^p$ for any $p > \frac{1}{2}$, so the bound $\frac{1}{2}$ in Theorem 2 is best possible. Again it may be true that all convex harmonic mappings are of class $h^{1/2}$.

The proof of Theorem 1 proceeds in three main steps. We may suppose without loss of generality that $f \in S_H$. Then $h'(\zeta) \neq 0$ for each $\zeta \in D$, and the function

$$F(z) = \frac{f\left(\frac{z+\zeta}{1+\bar{\zeta}z}\right) - f(\zeta)}{(1 - |\zeta|^2)h'(\zeta)} = H(z) + \overline{G(z)}$$

again belongs to S_H, with $H(z) = z + A_2(\zeta)z^2 + \cdots$, where

$$A_2(\zeta) = \frac{1}{2}\left\{(1 - |\zeta|^2)\frac{h''(\zeta)}{h'(\zeta)} - 2\bar{\zeta}\right\}.$$

Thus $|A_2(\zeta)| \leq A$ for all $\zeta \in D$, which implies that the function $\varphi(z) = \log h'(\zeta)$ has the property

$$(1 - |z|^2)|\varphi'(z)| \leq c, \qquad c = 2A + 2.$$

This says that φ is a *Bloch function*, with Bloch "norm"

$$\|\varphi\|_{\mathcal{B}} = \sup_{z \in \mathbb{D}}(1 - |z|^2)|\varphi'(z)| \leq c.$$

The second step is to deduce from the property $\|\varphi\|_{\mathcal{B}} \leq c$ that

$$M_p^p(r, h') \leq 2(1 + p^2c^2)(1 - r)^{-a}, \qquad 0 \leq r < 1,$$

where $a < p^2c^2$. The argument is technical and will not be given here.

The final step is to conclude that $h \in H^p$ for all $p < 1/c^2$. For this purpose we may appeal to the following lemma.

Lemma. *Let* $0 < p < 1$, *and suppose that* h *is a function analytic in* $\mathbb{D}$ *with the property*

$$\int_0^1 (1 - r)^{p-1}M_p^p(r, h')\,dr < \infty.$$

Then $h \in H^p$.

Since $p - 1 - a > p - 1 - p(pc^2) > -1$ under the assumption that $p < 1/c^2$, the integral in the lemma converges, and it follows that $h \in H^p$. In view of the inequality $|g'(z)| < |h'(z)|$, the same argument shows that $g \in H^p$ for $p < 1/c^2$, so that $f \in h^p$. This proves the theorem, at least for $p < (2A + 2)^{-2}$.

Proof of Lemma. The result was first established by T. M. Flett in 1972, but the following relatively simple proof is due to Mateljević and Pavlović [1]. First note that, for $0 \leq r < s < 1$,

$$M_p^p(s, h) - M_p^p(r, h) \leq \frac{1}{2\pi} \int_0^{2\pi} |h(se^{i\theta}) - h(re^{i\theta})|^p \, d\theta,$$

while

$$|h(se^{i\theta}) - h(re^{i\theta})| \leq \left| \int_r^s h'(\rho e^{i\theta}) e^{i\theta} \, d\rho \right|$$
$$\leq (s - r) \max_{r \leq \rho \leq s} |h'(\rho e^{i\theta})|,$$

so that

$$M_p^p(s, h) - M_p^p(r, h) \leq C(s - r)^p M_p^p(s, h')$$

by the Hardy–Littlewood maximal theorem. Now take $r_n = 1 - 2^{-n}$ for $n = 0, 1, 2, \ldots$, suppose that $h(0) = 0$, and apply the preceding inequality to see that

$$M_p^p(r_n, h) = \sum_{k=1}^n \left\{ M_p^p(r_k, h) - M_p^p(r_{k-1}, h) \right\}$$
$$\leq C \sum_{k=1}^n (r_k - r_{k-1})^p M_p^p(r_k, h')$$
$$= C \sum_{k=1}^n 2^{-kp} M_p^p(r_k, h')$$
$$\leq C \sum_{k=1}^n 2^{-k(p-1)} \int_{r_k}^{r_{k+1}} M_p^p(r, h') \, dr$$
$$\leq C \int_0^{r_{n+1}} (1 - r)^{p-1} M_p^p(r, h') \, dr,$$

where the constant C is not necessarily the same at each occurrence. Therefore, the convergence of the integral implies that $M_p^p(r_n, h)$ remains bounded as $n \to \infty$. Thus $h \in H^p$, as claimed. This proves the lemma. ∎

Further results on integral means are contained in the papers by Nowak [1] and Grigoryan and Nowak [1, 2].

In closing this chapter, one further topic should be mentioned. There are a number of interesting results on harmonic polynomials. A *harmonic polynomial* is a function of the form $f = h + \overline{g}$ where h and g are analytic polynomials. As in the case of analytic polynomials, one seeks to describe the regions of coefficients that correspond to univalent harmonic polynomials. The geometry of the image regions is of interest as well. These questions are discussed by Suffridge [1] and by Jahangiri, Morgan, and Suffridge [1]. Valence questions for harmonic polynomials have also been studied. See for instance Wilmshurst [1], Bshouty, Hengartner, and Suez [1], and Khavinson and Swiatek [1]. More general valence questions are considered by Lyzzaik [1, 2, 3, 5], Abu-Muhanna and Lyzzaik [1], and Neumann [1].

9

Minimal Surfaces

The final two chapters of this book will focus on a fundamental connection between harmonic mappings and minimal surfaces. Briefly, the connection arises from the fact that the Euclidean coordinates of a minimal surface are harmonic functions of isothermal parameters. The projection of a minimal graph onto its base plane therefore defines a harmonic mapping. Conversely, the harmonic mappings that lift to minimal surfaces have a simple description and the corresponding surfaces can be given by explicit formulas. This representation makes harmonic mappings an effective tool in the study of minimal surfaces.

9.1. Background in Surface Theory

Before turning to minimal surfaces, it may be useful to review some basic concepts in the classical differential geometry of surfaces. The proofs of many statements will be omitted, but the reader will find further details in texts on differential geometry such as Stoker [1] or Struik [1], or in the book by Osserman [3] on minimal surfaces.

A *surface* can be viewed informally as a two-dimensional set of points in three-dimensional Euclidean space $\mathbb{R}^3$. Formally, it is a two-dimensional manifold with a "smooth" structure. For most of our purposes it will suffice to regard a surface S as an equivalence class of differentiable mappings $X = \Phi(U)$ from a domain $D \subset \mathbb{R}^2$ onto a set $\Omega \subset \mathbb{R}^3$. Here $U = (u, v)$ and $X = (x, y, z)$ denote points in $\mathbb{R}^2$ and $\mathbb{R}^3$, respectively. Two representations are equivalent if they induce a diffeomorphism between parameter domains. The *Jacobian matrix* of any parametric representation Φ is

$$\begin{pmatrix} x_u & y_u & z_u \\ x_v & y_v & z_v \end{pmatrix}.$$

The surface S is said to be *regular at a point* if the Jacobian matrix has rank 2 there or, equivalently, if the two row vectors X_u and X_v are linearly independent. This means it is possible to solve locally for one of the coordinates x, y, or z in terms of the other two. The surface S is *regular* if it is regular at every point. Regularity is an intrinsic property of S, independent of the choice of parameters. Henceforth we will assume that S is regular.

A surface is said to be *embedded* in $\mathbb{R}^3$ if it has no self-intersections.

A *nonparametric surface* is one with the special form $z = f(x, y)$ or with x or y expressed in terms of the other two coordinates. Thus, a regular surface is *locally nonparametric* but need not be (globally) nonparametric. A nonparametric surface is also called a *graph*.

The *surface area* of S is defined by the integral

$$\iint_D \left\{ \left[\frac{\partial(x, y)}{\partial(u, v)} \right]^2 + \left[\frac{\partial(z, x)}{\partial(u, v)} \right]^2 + \left[\frac{\partial(y, z)}{\partial(u, v)} \right]^2 \right\}^{1/2} \, du \, dv,$$

where, for instance,

$$\frac{\partial(x, y)}{\partial(u, v)} = \begin{vmatrix} x_u & y_u \\ x_v & y_v \end{vmatrix} = x_u y_v - x_v y_u$$

is the Jacobian of x and y with respect to u and v. Again it can be shown that the definition of surface area is invariant under a diffeomorphic change of parameters.

A *curve* C on the surface S is defined as an equivalence class of mappings $X = \Phi(U(t))$ from a real interval $a \leq t \leq b$ to Ω. The *differential* of the mapping $X = \Phi(U)$ is

$$dX = X_u \, du + X_v \, dv.$$

The *arclength* s of a smooth curve C on S is found by integrating the square root of the differential form

$$ds^2 = \|dX\|^2 = dX \cdot dX = E \, du^2 + 2F \, du \, dv + G \, dv^2,$$

where

$$E = X_u \cdot X_u, \qquad F = X_u \cdot X_v, \qquad G = X_v \cdot X_v.$$

This is known as the *first fundamental form* of the surface. It is invariant under change of parameters although its individual coefficients are not. In particular, the arclength of C is well-defined, being independent of the choice of parameters.

Observe also that the first fundamental form, regarded as a quadratic form in the variables du and dv, is positive definite at each regular point of the surface. Consequently, $E > 0$, $G > 0$, and $EG - F^2 > 0$.

A *tangent vector* to the curve C at a point $X_0 = X(t_0)$ is

$$X'(t_0) = X_u u'(t_0) + X_v v'(t_0).$$

The *tangent plane* of S at X_0 is the set of all tangent vectors to curves on the surface through X_0, the two-dimensional subspace spanned by the (independent) vectors X_u and X_v. The cross product $X_u \times X_v$ is orthogonal to the tangent plane. When normalized to have unit length, it becomes the *unit normal vector*

$$\mathbf{n} = \frac{X_u \times X_v}{\|X_u \times X_v\|}$$

at the point X_0. Thus, the orientation of parameters assigns a local orientation to the surface. However, the surface may not be (globally) *orientable*; it may be impossible to assign a normal direction in a continuous and consistent manner over the whole surface. The Möbius strip and the Klein bottle are familiar examples of nonorientable surfaces.

The components of the cross product $X_u \times X_v$ are exactly the three Jacobians that appear in the formula for surface area. Thus,

$$\|X_u \times X_v\|^2 = \left[\frac{\partial(x, y)}{\partial(u, v)}\right]^2 + \left[\frac{\partial(z, x)}{\partial(u, v)}\right]^2 + \left[\frac{\partial(y, z)}{\partial(u, v)}\right]^2.$$

On the other hand, Lagrange's identity in vector analysis shows that

$$\|X_u \times X_v\|^2 = \|X_u\|^2 \|X_v\|^2 - (X_u \cdot X_v)^2 = EG - F^2.$$

The coefficients of the first fundamental form can therefore be used to calculate surface area.

Curvature is a second-order effect, requiring the assumption that the surface S has continuous second partial derivatives in its parametric representations. It will also be assumed that the curve C on S is regular ($U'(t) \neq \mathbf{0}$) and twice continuously differentiable. If C is parametrized in terms of arclength s, the tangent vector $\mathbf{T} = X'(s)$ has unit length and is called the *unit tangent vector*. The *curvature vector* $d\mathbf{T}/ds$ is orthogonal to $\mathbf{T}$. Its normal projection

$$k(\mathbf{T}) = \frac{d\mathbf{T}}{ds} \cdot \mathbf{n}$$

is called the *normal curvature* of S at X_0 in the direction $\mathbf{T}$. The normal curvature will be shown to depend only on the tangent direction $\mathbf{T}$ of the curve C at X_0. Intuitively, it measures the rate at which the surface is rising out of its tangent plane in a specified direction.

A more concrete way to define normal curvature is to consider only the normal sections of S. In other words, for each tangent direction $\mathbf{T}$ at X_0, let

C be the curve of intersection of the surface S with the plane through X_0 that contains both the normal vector $\mathbf{n}$ and the tangent vector $\mathbf{T}$. Then $d\mathbf{T}/ds$ is parallel to $\mathbf{n}$ and $k(\mathbf{T}) = \pm\|d\mathbf{T}/ds\|$, the sign depending on the choice of orientation of the surface.

The *principal curvatures* k_1 and k_2 of S at X_0 are the maximum and minimum of $k(\mathbf{T})$ as $\mathbf{T}$ ranges over all directions in the tangent space. The *mean curvature* of S at X_0 is the average value $H = \frac{1}{2}(k_1 + k_2)$, whereas the *Gauss curvature* is the product $K = k_1 k_2$. The beautiful *theorema egregium* of Gauss asserts that K is a "bending invariant," unchanged whenever the surface is deformed without stretching.

The mean curvature and the Gauss curvature can be computed in terms of surface invariants. By the chain rule, the unit tangent vector of the curve C is

$$\mathbf{T} = \frac{dX}{ds} = X_u \frac{du}{ds} + X_v \frac{dv}{ds},$$

and the curvature vector is

$$\frac{d\mathbf{T}}{ds} = X_{uu}\left(\frac{du}{ds}\right)^2 + 2X_{uv}\left(\frac{du}{ds}\right)\left(\frac{dv}{ds}\right) + X_{vv}\left(\frac{dv}{ds}\right)^2 + X_u\frac{d^2u}{ds^2} + X_v\frac{d^2v}{ds^2}.$$

But the normal vector $\mathbf{n}$ is orthogonal to both X_u and X_v, so the normal curvature is

$$k = \frac{d\mathbf{T}}{ds} \cdot \mathbf{n} = L\left(\frac{du}{ds}\right)^2 + 2M\left(\frac{du}{ds}\right)\left(\frac{dv}{ds}\right) + N\left(\frac{dv}{ds}\right)^2,$$

where

$$L = X_{uu} \cdot \mathbf{n}, \qquad M = X_{uv} \cdot \mathbf{n}, \qquad N = X_{vv} \cdot \mathbf{n}.$$

The differential form $L\,du^2 + 2M\,du\,dv + N\,dv^2$ is known as the *second fundamental form* of the surface. Like the first fundamental form, it is intrinsic to the surface, invariant under sense-preserving change of parameters.

Because the first fundamental form represents the arclength differential ds^2, the normal curvature may be expressed symbolically as a ratio of the two fundamental forms:

$$k = \frac{L\,du^2 + 2M\,du\,dv + N\,dv^2}{E\,du^2 + 2F\,du\,dv + G\,dv^2}.$$

More precisely, it may be viewed as a ratio of quadratic forms:

$$k = \frac{L\xi^2 + 2M\xi\eta + N\eta^2}{E\xi^2 + 2F\xi\eta + G\eta^2},$$

where $\xi = du/dt$ and $\eta = dv/dt$ are the derivatives of the parameters with respect to an arbitrary regular parametrization of the curve C. The homogeneity of this expression shows in particular that k depends only on the tangent direction $\mathbf{T}$ at the point X_0, not on any other properties of C.

The principal curvatures of S at X_0, or the maximum and minimum values of $k(\mathbf{T})$, can now be computed by the method of Lagrange multipliers. The maximum or minimum of

$$L\xi^2 + 2M\xi\eta + N\eta^2$$

under the constraint

$$E\xi^2 + 2F\xi\eta + G\eta^2 = 1$$

will be attained at a point where

$$L\xi + M\eta = \lambda(E\xi + F\eta),$$
$$M\xi + N\eta = \lambda(F\xi + G\eta).$$

Since $(\xi, \eta) \neq \mathbf{0}$, it follows that the determinant $|B - \lambda A| = 0$, where

$$A = \begin{pmatrix} E & F \\ F & G \end{pmatrix}, \qquad B = \begin{pmatrix} L & M \\ M & N \end{pmatrix}.$$

Multiplying the first equation by ξ and the second by η, then adding the two equations, we see that

$$L\xi^2 + 2M\xi\eta + N\eta^2 = \lambda(E\xi^2 + 2F\xi\eta + G\eta^2) = \lambda.$$

Thus, the two principal curvatures k_1 and k_2 are the eigenvalues λ, the two (real) roots of the quadratic equation $|B - \lambda A| = 0$, which takes the explicit form

$$(EG - F^2)\lambda^2 - (EN - 2FM + GL)\lambda + (LN - M^2) = 0,$$

whose two (real) roots are the principal curvatures k_1 and k_2. When this equation is written in the form

$$(EG - F^2)(\lambda - k_1)(\lambda - k_2) = 0,$$

it is apparent that the mean curvature is

$$H = \frac{1}{2}\frac{EN - 2FM + GL}{EG - F^2},$$

whereas the Gauss curvature is

$$K = \frac{LN - M^2}{EG - F^2}.$$

We now turn to minimal surfaces. A surface S is called a *minimal surface* if for each sufficiently small simple closed curve C on S the portion of S enclosed by C has the minimum area among all surfaces spanning C. Minimal surfaces can be constructed physically by dipping a loop of wire into soap solution. Because of surface tension, the resulting soap film will assume the shape that minimizes surface area.

A nonparametric minimal surface will be called a *minimal graph*. By a standard argument in the calculus of variations, a minimal graph $z = f(x, y)$ can be shown to satisfy the nonlinear partial differential equation

$$\left(1 + f_y{}^2\right) f_{xx} - 2 f_x f_y f_{xy} + \left(1 + f_x{}^2\right) f_{yy} = 0.$$

This equation has the elegant geometric interpretation that the mean curvature of the surface is everywhere equal to zero. Indeed, the coefficients of the first and second fundamental forms are easily computed in the nonparametric case, and the minimal surface equation is seen to reduce to $EN - 2FM + GL = 0$, or $H = 0$. It is convenient to take the identical vanishing of mean curvature as the *definition* of a minimal surface. As an immediate consequence, the Gauss curvature of a minimal surface is negative unless both principal curvatures vanish.

The simplest example of a minimal surface is of course the plane. Two other classical examples are the catenoid

$$z = \cosh^{-1} r, \qquad \text{where} \qquad r^2 = x^2 + y^2,$$

and the helicoid $z = \tan^{-1}(y/x)$, both shown in Figure 9.1. Aside from the plane, the catenoid is the only minimal surface of revolution, and the helicoid

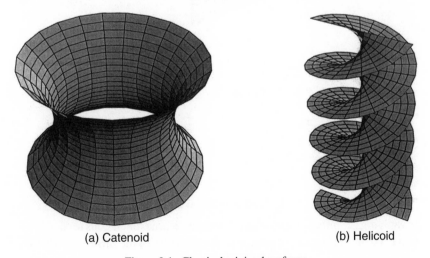

(a) Catenoid   (b) Helicoid

Figure 9.1. Classical minimal surfaces.

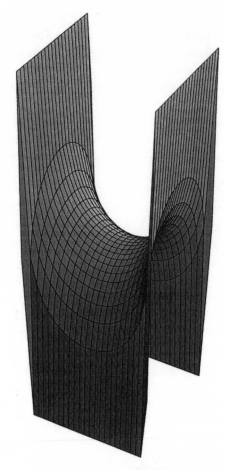

Figure 9.2. Scherk's surface.

is the only ruled minimal surface. The only minimal surface of translation $z = f(x) + g(y)$ is *Scherk's first surface*,

$$z = \log \frac{\cos y}{\cos x}, \qquad |x| < \frac{\pi}{2}, \qquad |y| < \frac{\pi}{2},$$

shown in Figure 9.2.

9.2. Isothermal Parameters

In studying the intrinsic properties of surfaces, it is advantageous to choose parameters that will reflect in some way the geometry of the surface. *Isothermal parameters* are those that preserve angles. In other words, the angle between a pair of curves in the parameter plane is equal to the angle between the corresponding pair of curves on the surface. Here the curves are oriented by their parametrizations, and the angle between them is understood to be the angle between their tangent vectors. As usual, this angle is chosen to lie between 0 and π.

To investigate the condition more closely, let $U = h(t)$ and $U = \tilde{h}(\tilde{t})$ be two curves in the parameter plane, intersecting at a point $U_0 = h(t_0) = \tilde{h}(\tilde{t}_0)$. The two tangent vectors are $h'(t_0) = (\xi, \eta)$ and $\tilde{h}'(\tilde{t}_0) = (\tilde{\xi}, \tilde{\eta})$, while the angle θ of intersection is given by

$$\|h'(t_0)\|\|\tilde{h}'(\tilde{t}_0)\| \cos \theta = h'(t_0) \cdot \tilde{h}'(\tilde{t}_0) = \xi\tilde{\xi} + \eta\tilde{\eta}.$$

On the other hand, the two curves lift to

$$X = \Phi(h(t)) = \psi(t) \qquad \text{and} \qquad \tilde{X} = \Phi(\tilde{h}(\tilde{t})) = \tilde{\psi}(\tilde{t})$$

on the surface, intersecting at a point $X_0 = \Phi(U_0)$ at an angle γ given by

$$\|\psi'(t_0)\|\|\tilde{\psi}'(\tilde{t}_0)\| \cos \gamma = \psi'(t_0) \cdot \tilde{\psi}'(\tilde{t}_0).$$

By the chain rule,

$$\psi'(t_0) = \Phi_u(U_0)\xi + \Phi_v(U_0)\eta$$

and

$$\tilde{\psi}'(\tilde{t}_0) = \Phi_u(U_0)\tilde{\xi} + \Phi_v(U_0)\tilde{\eta}.$$

Thus

$$\|\psi'(t_0)\|^2 = E\xi^2 + 2F\xi\eta + G\eta^2,$$
$$\|\tilde{\psi}'(\tilde{t}_0)\|^2 = E\tilde{\xi}^2 + 2F\tilde{\xi}\tilde{\eta} + G\tilde{\eta}^2,$$
$$\psi'(t_0) \cdot \tilde{\psi}'(\tilde{t}_0) = E\xi\tilde{\xi} + F(\xi\tilde{\eta} + \tilde{\xi}\eta) + G\eta\tilde{\eta},$$

where E, F, and G are the coefficients of the first fundamental form of the surface.

Suppose now that angles are preserved. In other words, suppose that the angle between a pair of curves meeting at the point U_0 in the parameter plane is always equal to the angle between the corresponding pair of curves through the point X_0 on the surface. Using the above expansions of the tangent vectors $\psi'(t_0)$ and $\tilde{\psi}'(\tilde{t}_0)$ to introduce E, F, and G into the formula for $\cos \gamma$, one finds that the equation $\cos \gamma = \cos \theta$ is equivalent to

$$\frac{E\xi\tilde{\xi} + F(\xi\tilde{\eta} + \tilde{\xi}\eta) + G\eta\tilde{\eta}}{\sqrt{E\xi^2 + 2F\xi\eta + G\eta^2}\sqrt{E\tilde{\xi}^2 + 2F\tilde{\xi}\tilde{\eta} + G\tilde{\eta}^2}} = \frac{\xi\tilde{\xi} + \eta\tilde{\eta}}{\sqrt{\xi^2 + \eta^2}\sqrt{\tilde{\xi}^2 + \tilde{\eta}^2}}.$$

Judicious choices of the tangent vectors (ξ, η) and $(\tilde{\xi}, \tilde{\eta})$ now lead to the conclusions that $F = 0$ and $E = G$. For instance, the choices $(\xi, \eta) = (1, 0)$ and $(\tilde{\xi}, \tilde{\eta}) = (0, 1)$ show that $F = 0$, whereas the choices $(\xi, \eta) = (1, 1)$ and $(\tilde{\xi}, \tilde{\eta}) = (1, -1)$ show that $E = G$. Conversely, it is evident that the relations $F = 0$ and $E = G$ imply $\cos \gamma = \cos \theta$, and so $\gamma = \theta$.

The conclusion is that angles between curves are preserved everywhere if and only if the first fundamental form has the structure

$$ds^2 = \lambda^2(du^2 + dv^2), \qquad \lambda = \lambda(u, v) > 0.$$

In this case the surface is said to be represented in terms of isothermal parameters.

It is true, but not at all obvious, that a regular surface can always be represented (locally) in terms of isothermal parameters. For *minimal* surfaces the proof is comparatively easy and may be found in Osserman [3]. For more general surfaces a proof can be based on quasiconformal mapping theory – specifically, on the existence of homeomorphic solutions to a Beltrami equation. (See Lehto [1], pp. 133–134, for further details.) An isothermal parametrization is unique up to precomposition with a conformal mapping or an anticonformal mapping (the complex conjugate of a conformal mapping).

We are now prepared to connect minimal surfaces with harmonic mappings. If a surface $X = \Phi(U)$ is expressed in terms of isothermal parameters, then, as we have just seen, $X_u \cdot X_v = 0$ and $X_u \cdot X_u = X_v \cdot X_v$. Further differentiations produce the relations

$$X_{uu} \cdot X_v + X_{uv} \cdot X_u = 0, \qquad X_{uv} \cdot X_v + X_{vv} \cdot X_u = 0;$$
$$X_{uu} \cdot X_u = X_{uv} \cdot X_v, \qquad X_{uv} \cdot X_u = X_{vv} \cdot X_v.$$

Combinations of these four equations show that the Laplacian $\Delta X = X_{uu} + X_{vv}$ is orthogonal to both X_u and X_v:

$$\Delta X \cdot X_u = \Delta X \cdot X_v = 0.$$

This means that ΔX is orthogonal to the tangent plane of the surface, so that $\|\Delta X\| = \pm \Delta X \cdot \mathbf{n} = \pm(L + N)$, where $\mathbf{n}$ is the unit normal vector, and $L = X_{uu} \cdot \mathbf{n}$ and $N = X_{vv} \cdot \mathbf{n}$ are coefficients in the second fundamental form.

Recall now that the mean curvature of the surface has the form

$$H = \frac{1}{2} \frac{EN - 2FM + GL}{EG - F^2}.$$

In isothermal parameters, $F = 0$ and $E = G = \lambda^2$, so the formula reduces to

$$H = \frac{1}{2\lambda^2}(L + N).$$

In view of the relation $\|\Delta X\| = \pm (L + N)$, this shows that $\Delta X = \mathbf{0}$ if and only if $H = 0$. But minimal surfaces are characterized by the vanishing of mean curvature. We have therefore proved the following theorem.

Theorem. *Let a regular surface S be expressed in terms of isothermal parameters. Then the position vector is a harmonic function of the parameters if and only if S is a minimal surface.*

Corollary. *If a nonparametric minimal surface is expressed in terms of isothermal parameters, the projection onto the base plane defines a harmonic mapping.*

To be more specific, suppose a nonparametric minimal surface $t = F(u, v)$ lies over a region Ω in the $uv-$ plane. Suppose it is represented by isothermal parameters (x, y) in a region D of the $xy-$ plane. Then the three coordinate functions $u = u(x, y), v = v(x, y),$ and $t = t(x, y) = F(u(x, y), v(x, y))$ are all harmonic. Because the mapping from D to the surface is injective and the surface is nonparametric, the projection $w = u + iv = f(z)$, where $z = x + iy$, defines a (univalent) harmonic mapping of D onto Ω.

9.3. Weierstrass–Enneper Representation

Again let a regular surface S have a parametric representation $X = \Phi(U)$, where with a slight change of notation $X = (x_1, x_2, x_3)$ and $U = (u, v)$ are points in $\mathbb{R}^3$ and $\mathbb{R}^2$, respectively. It is now advantageous to consider the complex variable $w = u + iv$ and to construct three complex-valued functions φ_k by the operation

$$2\frac{\partial X}{\partial w} = (\varphi_1, \varphi_2, \varphi_3).$$

In other words,

$$\varphi_k = 2\frac{\partial x_k}{\partial w} = \frac{\partial x_k}{\partial u} - i\frac{\partial x_k}{\partial v}, \qquad k = 1, 2, 3.$$

Recalling the definitions of E, F, and G in the first fundamental form of S, one finds by direct calculation that

$$\sum_{k=1}^{3} \varphi_k{}^2 = E - G - 2iF, \qquad \sum_{k=1}^{3} |\varphi_k|^2 = E + G. \qquad (1)$$

Suppose now that the parametric representation of S is isothermal. Then $F = 0$ and $E = G > 0$, so

$$\sum_{k=1}^{3} \varphi_k(w)^2 = 0, \qquad \sum_{k=1}^{3} |\varphi_k(w)|^2 > 0. \qquad (2)$$

If S is a minimal surface, the coordinates x_k are harmonic and, hence, the functions φ_k are analytic. Consequently, to every regular minimal surface there correspond three analytic functions φ_k with the properties (2). Furthermore, the process is reversible and the converse is true. In other words, each triple of analytic functions satisfying (2) will generate a regular minimal surface. Let us formulate the result more precisely as a theorem.

Theorem 1. *Let $X = \Phi(U)$ be an isothermal parametrization of a regular minimal surface and let $w = u + iv$. Then the functions*

$$\varphi_k = 2\frac{\partial x_k}{\partial w}, \qquad k = 1, 2, 3,$$

are analytic and have the properties (2). Conversely, let $\{\varphi_1, \varphi_2, \varphi_3\}$ be an arbitrary triple of functions analytic in a simply connected region, satisfying (2). Then the functions

$$x_k = \operatorname{Re}\left\{\int \varphi_k(w)\, dw\right\}, \qquad k = 1, 2, 3, \qquad (3)$$

give an isothermal parametrization of a regular minimal surface.

Proof. We have already established the first statement. Only the converse remains to be proved. The functions $x_k = x_k(w)$ defined by equations (3) are harmonic and give a parametric representation of a surface S. Noting that

$$\frac{\partial x_k}{\partial w} = \frac{1}{2}\varphi_k, \qquad k = 1, 2, 3,$$

and referring to the formulas (1), we see in view of properties (2) that $E = G > 0$ and $F = 0$, so S is a regular surface and the parametric representation is isothermal. Finally, since the coordinates are harmonic functions of isothermal parameters, we conclude from the theorem of Section 9.2 that S is a minimal surface. It need not be embedded, however. ∎

The simplest choice of analytic functions φ_k satisfying (2) is $\varphi_1(w) = 1$, $\varphi_2(w) = -i$, $\varphi_3(w) = 0$. Then the formulas (3) reduce to $x_1 = u$, $x_2 = v$, and $x_3 = 0$. The minimal surface is the coordinate plane itself, and the parametric representation is obviously isothermal. Some examples of greater interest will be developed in the next section.

In the 1860s, Karl Weierstrass and Alfred Enneper arrived independently at the foregoing representation of a minimal surface in terms of a triple of analytic functions, and they made the further important discovery that all such triples can be described explicitly. The combination of these two results is now known as the *Weierstrass–Enneper representation* of a minimal surface. The following lemma describes the relevant triples of analytic functions.

Lemma. *Let p be an analytic function and q a meromorphic function in some domain $D \subset \mathbb{C}$. Suppose that p has a zero of order at least $2m$ wherever q has a pole of order m. Then the functions*

$$\varphi_1 = p(1 + q^2), \qquad \varphi_2 = -ip(1 - q^2), \qquad \varphi_3 = -2ipq \qquad (4)$$

are analytic in D and have the property

$$\varphi_1{}^2 + \varphi_2{}^2 + \varphi_3{}^2 = 0. \qquad (5)$$

Conversely, every ordered triple of functions φ_1, φ_2, φ_3 analytic in D with the property (5) has the structure (4), unless $\varphi_2 = i\varphi_1$. The representation is unique.

Proof. It is easy to verify that every triple with the structure (4) satisfies (5). The condition on the zeros of p ensures that the functions φ_k are analytic. Conversely, let φ_k be any analytic functions with the property (5). If $\varphi_2 \neq i\varphi_1$, we may define

$$p = \frac{1}{2}(\varphi_1 + i\varphi_2), \qquad q = \frac{i\varphi_3}{\varphi_1 + i\varphi_2}, \qquad (6)$$

so that $2pq = i\varphi_3$. To verify the other two parts of (4), rewrite the condition (5) as

$$(\varphi_1 + i\varphi_2)(\varphi_1 - i\varphi_2) = -\varphi_3{}^2,$$

and deduce that

$$pq^2 = \frac{1}{2}(\varphi_1 - i\varphi_2).$$

Hence,

$$p(1 + q^2) = \varphi_1, \qquad p(1 - q^2) = i\varphi_2.$$

The uniqueness of the representation is clear, since the equations (4) can be solved for p and q as in (6). If $\varphi_2 = i\varphi_1$, then $\varphi_1^2 + \varphi_2^2 = 0$, and it follows that $\varphi_3 = 0$. The representation (4) then follows with the uniquely determined choices $p = \varphi_1$ and $q = 0$. In this degenerate case, the corresponding minimal surface is a horizontal plane. ∎

The lemma allows Theorem 1 to be recast in a more useful way as follows.

Theorem 2 (Weierstrass–Enneper Representation). *Every regular minimal surface has locally an isothermal parametric representation of the form*

$$\begin{cases} x_1 = \text{Re}\left\{ \int p(1 + q^2)\,dw \right\}, \\ x_2 = \text{Im}\left\{ \int p(1 - q^2)\,dw \right\}, \\ x_3 = 2\,\text{Im}\left\{ \int pq\,dw \right\} \end{cases} \tag{7}$$

in some domain $D \subset \mathbb{C}$, where p is analytic and q is meromorphic in D, with p vanishing only at the poles (if any) of q and having a zero of precise order $2m$ wherever q has a pole of order m. Conversely, each such pair of functions p and q analytic and meromorphic, respectively, in a simply connected domain D generates through the formulas (7) an isothermal parametric representation of a regular minimal surface.

Note that the more stringent restriction on the zeros of p implies that the analytic functions φ_1, φ_2, and φ_3 have no common zeros, since a simple calculation gives

$$\sum_{k=1}^{3} |\varphi_k|^2 = 2|p|^2(1 + |q|^2)^2 > 0.$$

This guarantees the regularity of the associated minimal surface. If the underlying domain D is not simply connected, the integrals that define the parametrization may be multiple-valued.

The function q in the Weierstrass–Enneper representation has a beautiful geometric interpretation. Recall that the *Gauss map* of a surface sends each point X to the point on the unit sphere that corresponds to the unit normal vector of the surface at X. It will now be shown that $-i/q(w)$ is the stereographic projection of the image of $X(w)$ under the Gauss map. In particular, the normal direction depends only on q and not on p.

For the proof, it is necessary to calculate the unit normal vector in terms of the parametrization (7) of the surface. Observe first that, because isothermal parameters are in use, the vectors X_u and X_v are orthogonal and so their cross product has norm

$$\|X_u \times X_v\| = \|X_u\|\|X_v\| = \lambda^2 = \frac{1}{2}\sum_{k=1}^{3}|\varphi_k|^2,$$

or

$$\|X_u \times X_v\| = |p|^2(1+|q|^2)^2. \tag{8}$$

To compute the cross product itself in terms of p and q, it is convenient to invoke the relation

$$X_u \times X_v = \operatorname{Im}\{(\varphi_2\overline{\varphi_3},\ \varphi_3\overline{\varphi_1},\ \varphi_1\overline{\varphi_2})\},$$

which is verified directly from the definition $\varphi_k = 2\,\partial x_k/\partial w$. Substitution of the formulas (4) leads after brief calculation to the simple expression

$$X_u \times X_v = -|p|^2(1+|q|^2)(2\operatorname{Im}\{q\},\ 2\operatorname{Re}\{q\},\ |q|^2-1).$$

Note that this again gives formula (8) for the norm of the cross product. The unit normal vector is now found to be

$$\mathbf{n} = \frac{X_u \times X_v}{\|X_u \times X_v\|} = -\frac{1}{1+|q|^2}(2\operatorname{Im}\{q\},\ 2\operatorname{Re}\{q\},\ |q|^2-1). \tag{9}$$

On the other hand, it happens that the inverse stereographic projection of a point $z = x + iy$ in the complex plane onto the unit sphere is precisely

$$\frac{1}{1+|z|^2}(2x, 2y, |z|^2-1)$$

(see, for instance, Ahlfors [3], p. 18). Comparison with formula (9) for the unit normal vector completes the identification of $-i/q(w)$ as the image of $X(w)$ under the Gauss map followed by stereographic projection.

9.4. Some Examples

The Weierstrass–Enneper representation allows the explicit construction of a wide variety of minimal surfaces. One has only to choose an analytic function p and a meromorphic function q whose zeros and poles are related as in

Theorem 2 of the previous section, and to perform the required integrations to obtain a regular minimal surface represented by isothermal parameters. In principle, every minimal surface can be obtained in this way by suitable choice of p and q. The construction will now be illustrated with some of the most common examples.

(i) *The Plane.* The choice of constant functions p and q gives rise to a plane represented by isothermal parameters.

(ii) *The Catenoid.* Take $p(w) \equiv 1$ and $q(w) = i/w$ in the domain $D = \mathbb{C} \setminus \{0\}$, the complex plane punctured at the origin. Then

$$x_1 = \mathrm{Re}\{\int p(1 + q^2)\,dw\} = \left(r + \tfrac{1}{r}\right)\cos\theta,$$
$$x_2 = \mathrm{Im}\{\int p(1 - q^2)dw\} = \left(r + \tfrac{1}{r}\right)\sin\theta,$$
$$x_3 = 2\,\mathrm{Im}\{\int pq\,dw\} = 2\log r,$$

where $w = re^{i\theta}$. With $\rho = \log r$, the equations take the form

$$x_1 = 2\cosh\rho\,\cos\theta, \qquad x_2 = 2\cosh\rho\,\sin\theta, \qquad x_3 = 2\rho,$$

which is easily recognized as a parametric representation of a catenoid.

(iii) *The Helicoid.* Again let $D = \mathbb{C} \setminus \{0\}$ and take $p(w) \equiv 1, q(w) = 1/w$. Then

$$x_1 = \left(r - \frac{1}{r}\right)\cos\theta, \qquad x_2 = \left(r - \frac{1}{r}\right)\sin\theta, \qquad x_3 = 2\theta,$$

where $w = re^{i\theta}$. These are the equations of a helicoid. Strictly speaking, a branch cut is needed (say, along the positive real axis) to make the formula for x_3 single-valued. The equations then represent one turn of the helicoid.

(iv) *Enneper's Surface.* Take $D = \mathbb{C}$, $p(w) \equiv 1$, and $q(w) = iw$. Simple integrations then produce the formulas

$$x_1 = \mathrm{Re}\left\{w - \frac{1}{3}w^3\right\} = u + uv^2 - \frac{1}{3}u^3,$$
$$x_2 = \mathrm{Im}\left\{w + \frac{1}{3}w^3\right\} = v + u^2v - \frac{1}{3}v^3,$$
$$x_3 = \mathrm{Re}\{w^2\} = u^2 - v^2,$$

where $w = u + iv$. The equations define what is known as *Enneper's surface*. Figure 9.3 shows the parts of the surface that correspond to the unit disk $|w| < 1$ and to the disk $|w| < 2$. As the parametric disk

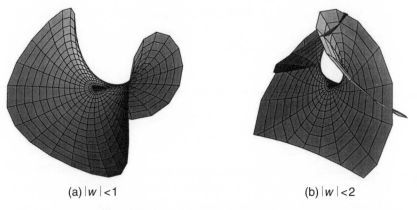

(a) $|w| < 1$ (b) $|w| < 2$

Figure 9.3. Enneper's surface.

expands, the surface overlaps itself and eventually undergoes self-intersections.

(v) *Scherk's First Surface.* Let D be the unit disk and take $p(w) = 2/(1 - w^4)$, $q(w) = iw$. The relevant integrals are

$$\int_0^w p(1 + q^2)\, dw = \int_0^w \frac{2}{1 + w^2}\, dw = 2\tan^{-1} w = -i \log \frac{1 + iw}{1 - iw},$$

$$\int_0^w p(1 - q^2)\, dw = \int_0^w \frac{2}{1 - w^2}\, dw = \log \frac{1 + w}{1 - w},$$

$$\int_0^w pq\, dw = \int_0^w \frac{2iw}{1 - w^4}\, dw = \frac{i}{2} \log \frac{1 + w^2}{1 - w^2}.$$

Thus, the minimal surface has the equations

$$x_1 = \arg\left\{\frac{1 + iw}{1 - iw}\right\}, \qquad x_2 = \arg\left\{\frac{1 + w}{1 - w}\right\}, \qquad x_3 = \log\left|\frac{1 + w^2}{1 - w^2}\right|.$$

Note that x_1 and x_2 lie in the interval $(-\pi/2, \pi/2)$. With the help of the formulas

$$\frac{1 + iw}{1 - iw} = \frac{1}{|1 - iw|^2}[(1 - |w|^2) + 2iu]$$

and

$$\frac{1 + w}{1 - w} = \frac{1}{|1 - w|^2}[(1 - |w|^2) + 2iv],$$

where $w = u + iv$, a calculation now shows that

$$\left(\frac{\cos x_2}{\cos x_1}\right)^2 = \frac{1 + 2(u^2 - v^2) + (u^2 + v^2)^2}{1 - 2(u^2 - v^2) + (u^2 + v^2)^2} = \left|\frac{1 + w^2}{1 - w^2}\right|^2,$$

so that

$$x_3 = \log \frac{\cos x_2}{\cos x_1},$$

the defining equation of Scherk's surface.

9.5. Historical Notes

The theory of minimal surfaces originated with Euler and Lagrange, who considered the area-minimizing problem as an application of the calculus of variations. The partial differential equation for a minimal surface dates back to work of J. L. Lagrange published in 1762. The geometric interpretation as vanishing mean curvature is due to J. B. Meusnier in 1776. The catenoid, helicoid, and plane were the only known examples of minimal surfaces with explicit parametric representations until the year 1832, when Heinrich F. Scherk described his first surface. In two papers published in 1832 and 1835, Scherk described a total of five new minimal surfaces, including his "saddle-tower" surface that is discussed in Chapter 10. Many more examples were then discovered as the subject enjoyed a period of rapid development through the middle of the Nineteenth Century. Karl Weierstrass and Alfred Enneper arrived at their representation formula independently in 1866 and 1864, respectively. Harmonic mappings were not seriously investigated until the 1920s, when Tibor Radó studied them in connection with Plateau's problem.

The treatise by Johannes Nitsche [7] contains a thorough and scholarly account of the historical development of minimal surfaces, including a discussion of harmonic mappings. For further information on minimal surfaces, the reader may consult Osserman [3, 4], Hoffman [1], and the book by Dierkes, Hildebrandt, Küster, and Wohlrab [1].

10

Curvature of Minimal Surfaces

Continuing the development of the previous chapter, we now take a closer look at the connection between harmonic mappings and nonparametric minimal surfaces with a view to obtaining sharp bounds for the Gauss curvature. Our first task is to derive a useful formula for Gauss curvature in terms of isothermal parameters. We then specialize the formula to minimal surfaces and express the curvature in terms of the underlying harmonic mapping. This allows us to estimate the curvature of a minimal surface by appeal to Heinz' lemma and related results about harmonic mappings.

10.1. Gauss Curvature

In the previous chapter we found that the Gauss curvature of a surface has the expression

$$K = \frac{LN - M^2}{EG - F^2},$$

where $E = X_u \cdot X_u$, $F = X_u \cdot X_v$, $G = X_v \cdot X_v$ are the coefficients of the first fundamental form; and $L = X_{uu} \cdot \mathbf{n}$, $M = X_{uv} \cdot \mathbf{n}$, $N = X_{vv} \cdot \mathbf{n}$ are those of the second fundamental form. We shall now show that if isothermal parameters are chosen, so that $F = 0$, $E = G = \lambda^2$, and the first fundamental form reduces to

$$ds^2 = \lambda^2(du^2 + dv^2), \qquad \lambda = \lambda(u, v) > 0,$$

then Gauss curvature has the elegant expression

$$K = -\frac{1}{\lambda^2}\Delta(\log \lambda),$$

where $\Delta = \partial^2/\partial u^2 + \partial^2/\partial v^2$ denotes the Laplacian. This will show in particular that Gauss curvature depends only on the coefficients of first fundamental form and therefore remains invariant when the surface is deformed without stretching, which is the celebrated *theorema egregium* of Gauss.

Recalling that the unit normal vector is

$$\mathbf{n} = \frac{X_u \times X_v}{\|X_u \times X_v\|} = \frac{1}{\sqrt{EG - F^2}} \left\{ \frac{\partial(y, z)}{\partial(u, v)} \mathbf{i} + \frac{\partial(z, x)}{\partial(u, v)} \mathbf{j} + \frac{\partial(x, y)}{\partial(u, v)} \mathbf{k} \right\},$$

we see that

$$L = X_{uu} \cdot \mathbf{n} = \frac{1}{\sqrt{EG - F^2}} \begin{vmatrix} x_{uu} & y_{uu} & z_{uu} \\ x_u & y_u & z_u \\ x_v & y_v & z_v \end{vmatrix},$$

$$M = X_{uv} \cdot \mathbf{n} = \frac{1}{\sqrt{EG - F^2}} \begin{vmatrix} x_{uv} & y_{uv} & z_{uv} \\ x_u & y_u & z_u \\ x_v & y_v & z_v \end{vmatrix},$$

$$N = X_{vv} \cdot \mathbf{n} = \frac{1}{\sqrt{EG - F^2}} \begin{vmatrix} x_{vv} & y_{vv} & z_{vv} \\ x_u & y_u & z_u \\ x_v & y_v & z_v \end{vmatrix}.$$

Consequently,

$$(EG - F^2)LN = \begin{vmatrix} X_{uu} \cdot X_{vv} & X_{uu} \cdot X_u & X_{uu} \cdot X_v \\ X_u \cdot X_{vv} & X_u \cdot X_u & X_u \cdot X_v \\ X_v \cdot X_{vv} & X_v \cdot X_u & X_v \cdot X_v \end{vmatrix}$$

$$= \begin{vmatrix} X_{uu} \cdot X_{vv} & \frac{1}{2}E_u & X_{uu} \cdot X_v \\ X_u \cdot X_{vv} & E & F \\ \frac{1}{2}E_v & F & G \end{vmatrix},$$

$$(EG - F^2)M^2 = \begin{vmatrix} X_{uv} \cdot X_{uv} & X_{uv} \cdot X_u & X_{uv} \cdot X_v \\ X_u \cdot X_{uv} & X_u \cdot X_u & X_u \cdot X_v \\ X_v \cdot X_{uv} & X_v \cdot X_u & X_v \cdot X_v \end{vmatrix}$$

$$= \begin{vmatrix} X_{uv} \cdot X_{uv} & \frac{1}{2}E_v & \frac{1}{2}G_u \\ \frac{1}{2}E_v & E & F \\ \frac{1}{2}G_u & F & G \end{vmatrix}.$$

Assume now that the surface is expressed in terms of isothermal parameters, so that $E = G$ and $F = 0$. Expanding each of the determinants along the first row, we find

$$LN = X_{uu} \cdot X_{vv} - \frac{E_u}{2E} X_{vv} \cdot X_u - \frac{E_v}{2E} X_{uu} \cdot X_v$$

and

$$M^2 = X_{uv} \cdot X_{uv} - \frac{E_v{}^2}{4E} - \frac{E_u{}^2}{4E}.$$

On the other hand, since

$$X_{uu} \cdot X_v + X_u \cdot X_{vu} = F_u = 0,$$

it is clear that

$$X_{uu} \cdot X_v = -X_{uv} \cdot X_u = -\frac{1}{2} E_v.$$

Similarly, the relation

$$X_{uv} \cdot X_v + X_u \cdot X_{vv} = F_v = 0$$

shows that

$$X_{vv} \cdot X_u = -X_{vu} \cdot X_v = -\frac{1}{2} G_u = -\frac{1}{2} E_u.$$

Finally, the identity

$$X_{uu} \cdot X_{vv} - X_{uv} \cdot X_{uv} = -\frac{1}{2} E_{vv} + F_{uv} - \frac{1}{2} G_{uu}$$

is easily verified by straightforward calculation. Thus, for isothermal parameters with $E = G = \lambda^2$ and $F = 0$, we arrive at the expression

$$LN - M^2 = -\frac{1}{2}(E_{uu} + E_{vv}) + \frac{E_u{}^2}{2E} + \frac{E_v{}^2}{2E}$$
$$= \lambda_u{}^2 - \lambda\lambda_{uu} + \lambda_v{}^2 - \lambda\lambda_{vv} = -\lambda^2 \Delta(\log \lambda),$$

which is the desired result because $EG - F^2 = \lambda^4$.

10.2. Minimal Graphs and Harmonic Mappings

It was shown in the previous chapter that when a minimal surface is represented by isothermal parameters, its three coordinate functions are harmonic. As a consequence, the projection of a minimal graph to its base plane is a harmonic mapping. Our object is now to characterize the harmonic mappings obtained in this way and to show how they lift to minimal surfaces.

Consider a regular minimal graph

$$S = \{(u, v, F(u, v)) : u + iv \in \Omega\}$$

over a simply connected domain $\Omega \subset \mathbb{C}$ containing the origin. Suppose that Ω is not the whole plane. (It will be shown later that the only minimal graphs that

extend over the entire plane are themselves planes, a theorem of S. Bernstein.)
In view of the Weierstrass–Enneper representation, as developed in Section
9.3, the surface has a reparametrization by isothermal parameters $z = x + iy$
in the unit disk $\mathbb{D}$ so that

$$u = \text{Re}\left\{ \int_0^z \varphi_1(\zeta)\,d\zeta \right\}, \qquad v = \text{Re}\left\{ \int_0^z \varphi_2(\zeta)\,d\zeta \right\},$$

$$F(u, v) = \text{Re}\left\{ \int_0^z \varphi_3(\zeta)\,d\zeta \right\}, \qquad z \in \mathbb{D},$$

where the functions φ_k are analytic in $\mathbb{D}$ and satisfy

$$\sum_{k=1}^3 \varphi_k(z)^2 = 0, \qquad \sum_{k=1}^3 |\varphi_k(z)|^2 > 0.$$

There is no loss of generality in supposing that z ranges over the unit disk,
because any other isothermal representation can be precomposed with a conformal map from the disk whose existence is guaranteed by the Riemann
mapping theorem. The functions φ_k may be expressed in the form

$$\varphi_1 = p(1 + q^2), \qquad \varphi_2 = -ip(1 - q^2), \qquad \varphi_3 = -2ipq,$$

where p is analytic and q is meromorphic in $\mathbb{D}$, with p nonvanishing except
for a zero of order $2m$ wherever q has a pole of order m. In terms of p and q,
the Weierstrass–Enneper representation is

$$u = \text{Re}\left\{ \int_0^z p(1 + q^2)\,d\zeta \right\}, \qquad v = \text{Im}\left\{ \int_0^z p(1 - q^2)\,d\zeta \right\},$$

$$F(u, v) = 2\,\text{Im}\left\{ \int_0^z pq\,d\zeta \right\}, \qquad z \in \mathbb{D}.$$

Now let $w = u + iv$ and let $w = f(z)$ denote the projection of S onto its
base plane:

$$f(z) = \text{Re}\left\{ \int_0^z \varphi_1(\zeta)\,d\zeta \right\} + i\,\text{Re}\left\{ \int_0^z \varphi_2(\zeta)\,d\zeta \right\}.$$

Then f is a harmonic mapping of $\mathbb{D}$ onto Ω with $f(0) = 0$. Let

$$f = h + \overline{g}, \qquad h(0) = g(0) = 0,$$

be the canonical decomposition of f, where h and g are analytic in $\mathbb{D}$. Differentiation leads to the formulas

$$h' = \frac{1}{2}(\varphi_1 + i\varphi_2), \qquad g' = \frac{1}{2}(\varphi_1 - i\varphi_2),$$

or

$$\varphi_1 = h' + g', \qquad \varphi_2 = -i(h' - g').$$

Hence, a simple calculation gives

$$\varphi_3{}^2 = -\varphi_1{}^2 - \varphi_2{}^2 = -4h'g' = -4\omega h'^2,$$

where $\omega = g'/h'$ is the dilatation of f. This shows that $\omega = -\frac{1}{4}\varphi_3{}^2/h'^2$ is the *square* of a meromorphic function. In other words, the harmonic mappings that result from projection of minimal graphs have dilatations with single-valued square roots. If f is sense-preserving, this is eqivalent to saying that its dilatation function ω has no zeros of odd order.

The formula for ω is further illuminated when the Weierstrass–Enneper functions p and q are introduced. Then

$$f(z) = \mathrm{Re}\left\{ \int_0^z p(1 + q^2)\,d\zeta \right\} + i\,\mathrm{Im}\left\{ \int_0^z p(1 - q^2)\,d\zeta \right\}$$

and a similar calculation shows that $h' = p$ and $g' = pq^2$, which gives the elegant expression $\omega = q^2$ for the dilatation of the projected harmonic mapping f. In particular, f is sense-preserving if and only if q is analytic and $|q(z)| < 1$ in $\mathbb{D}$. In view of the remarks at the end of Section 9.3, the relation $\omega = q^2$ also identifies $-i/\sqrt{w}$ as the stereographic projection of the Gauss map of the corresponding minimal surface.

The problem now arises to give a full description of the harmonic mappings that are projections of minimal surfaces. In other words, what properties of a harmonic mapping are necessary and sufficient for it to lift to a minimal graph expressed by isothermal parameters? A necessary condition, as we have just shown, is that the dilatation of the harmonic mapping is the square of a meromorphic function. Surprisingly, the condition is also sufficient. To verify this assertion, we may suppose without loss of generality that the mapping is sense-preserving.

Theorem. *If a minimal graph*

$$\{(u, v, F(u, v)) : u + iv \in \Omega\}$$

is parametrized by sense-preserving isothermal parameters $z = x + iy \in \mathbb{D}$, the projection onto its base plane defines a harmonic mapping $w = u + iv = f(z)$ of $\mathbb{D}$ onto Ω whose dilatation is the square of an analytic function. Conversely, if $f = h + \overline{g}$ is a sense-preserving harmonic mapping of $\mathbb{D}$ onto some domain Ω with dilatation $\omega = q^2$ for some function q analytic in $\mathbb{D}$, *then the formulas*

$$u = \text{Re}\{f(z)\}, \qquad v = \text{Im}\{f(z)\}, \qquad t = 2\text{Im}\left\{\int_0^z q(\zeta)h'(\zeta)\,d\zeta\right\} \quad (1)$$

define by isothermal parameters a minimal graph whose projection is f. Except for the choice of sign and an arbitrary additive constant in the third coordinate function, this is the only such surface.

Proof. The necessity of the condition $\omega = q^2$ has already been proved. For the converse it need only be shown, in view of the theorem in Section 9.2, that the surface defined by equations (1) is represented by harmonic functions of isothermal parameters. According to the discussion in Section 9.3, this is equivalent to showing that each of the derivatives $\partial u/\partial z$, $\partial v/\partial z$, and $\partial t/\partial z$ is analytic, and that

$$\left(\frac{\partial u}{\partial z}\right)^2 + \left(\frac{\partial v}{\partial z}\right)^2 + \left(\frac{\partial t}{\partial z}\right)^2 = 0. \tag{2}$$

But direct calculations lead to the expressions

$$\frac{\partial u}{\partial z} = \frac{1}{2}(h' + g'), \qquad \frac{\partial v}{\partial z} = \frac{1}{2i}(h' - g'), \qquad \frac{\partial t}{\partial z} = -iqh',$$

where $q^2 = \omega = g'/h'$. A further calculation now gives the desired conclusion (2). Since f is univalent by hypothesis, the given surface is immediately seen to be a graph: the third coordinate t is actually just a function of u and v.

To verify the uniqueness assertion, let

$$u = \text{Re}\{f(z)\}, \qquad v = \text{Im}\{f(z)\}, \qquad t = k(z)$$

represent some other minimal surface in isothermal parameters. Then k is harmonic, so that $\partial k/\partial z$ is analytic. Since the representation is isothermal, relation (2) must hold, as shown at the beginning of Section 9.3. This implies that

$$\left(\frac{\partial k}{\partial z}\right)^2 = -\left(\frac{\partial u}{\partial z}\right)^2 - \left(\frac{\partial v}{\partial z}\right)^2 = -h'g' = -q^2h'^2,$$

so that $\partial k/\partial z = \pm iqh'$. But the real-valued harmonic function k has a unique representation $k = \psi + \overline{\psi} = 2\text{Re}\{\psi\}$ for some analytic function ψ. Since $\psi' = \pm iqh'$, it follows that

$$\psi(z) = \pm i \int_0^z qh'\,d\zeta + C$$

for some complex constant C, which proves the uniqueness. ∎

Two examples will now be given to illustrate the process of lifting a harmonic mapping to a minimal surface. First consider the function $f(z) = z - \frac{1}{3}\bar{z}^3$, which provides a harmonic mapping of the unit disk $\mathbb{D}$ onto the region Ω inside a hypocycloid of four cusps inscribed in the circle $|w| = 4/3$ (see Section 1.1). Here $h(z) = z$ and $g(z) = -\frac{1}{3}z^3$, so the dilatation of f is $\omega(z) = -z^2$. Because ω is the square of an analytic function, the theorem says that f lifts to the minimal surface defined by the equations

$$u = \text{Re}\{f(z)\} = x + xy^2 - \frac{1}{3}x^3, \qquad v = \text{Im}\{f(z)\} = y + x^2y - \frac{1}{3}y^3,$$

$$t = 2\,\text{Im}\left\{\int_0^z \sqrt{\omega}\, h'\, d\zeta\right\} = 2\,\text{Im}\left\{-i\int_0^z \zeta\, d\zeta\right\} = \text{Re}\{z^2\} = x^2 - y^2,$$

where $z = x + iy$ ranges over $\mathbb{D}$. This is a nonparametric portion of Enneper's surface, as presented in Section 9.4. The complete surface is obtained as z ranges over the whole plane.

For a second example, consider the function

$$f(z) = \frac{1+i}{\pi\sqrt{2}} \sum_{k=0}^{3} i^k \arg\left\{\frac{z - i^k}{z - i^{k+1}}\right\},$$

which maps $\mathbb{D}$ harmonically onto the square region

$$\Omega = \left\{w = u + iv : -\frac{1}{\sqrt{2}} < u < \frac{1}{\sqrt{2}}, -\frac{1}{\sqrt{2}} < v < \frac{1}{\sqrt{2}}\right\},$$

inscribed in the unit circle. Here f has a piecewise constant boundary function

$$f(e^{i\theta}) = \frac{1}{\sqrt{2}}(1+i)i^k, \qquad \frac{k\pi}{2} < \theta < \frac{(k+1)\pi}{2},$$

for $k = 0, 1, 2, 3$. Note that the construction is the same as in Section 4.2, with minor modifications. The mapping function has a canonical decomposition $f = h + \bar{g}$, where

$$h(z) = \frac{1-i}{2\sqrt{2}\pi} \sum_{k=0}^{3} i^k \log\left\{\frac{z - i^k}{z - i^{k+1}}\right\}$$

and

$$g(z) = -\frac{1+i}{2\sqrt{2}\pi} \sum_{k=0}^{3} (-i)^k \log\left\{\frac{z - i^k}{z - i^{k+1}}\right\}.$$

Differentiation produces the expressions

$$h'(z) = \frac{\sqrt{2}}{\pi}\left(\frac{1}{z^2 - 1} - \frac{1}{z^2 + 1}\right) = \frac{2\sqrt{2}}{\pi}\frac{1}{z^4 - 1}$$

and

$$g'(z) = -\frac{\sqrt{2}}{\pi} \left(\frac{1}{z^2 - 1} + \frac{1}{z^2 + 1} \right) = -\frac{2\sqrt{2}}{\pi} \frac{z^2}{z^4 - 1}.$$

Thus, f has dilatation $\omega(z) = g'(z)/h'(z) = -z^2$, and f lifts to the minimal surface with parametric equations

$$u = \text{Re}\{f(z)\} = \text{Re}\{h(z) + g(z)\},$$
$$v = \text{Im}\{f(z)\} = \text{Im}\{h(z) - g(z)\},$$
$$t = -2\,\text{Re}\left\{ \int_0^z \zeta h'(\zeta)\, d\zeta \right\}.$$

After further calculation, these formulas reduce to

$$u = \frac{\sqrt{2}}{\pi} \arg\left\{ \frac{1 - iz}{1 + iz} \right\}, \qquad v = \frac{\sqrt{2}}{\pi} \arg\left\{ \frac{1 - z}{1 + z} \right\},$$
$$t = \frac{\sqrt{2}}{\pi} \log\left| \frac{1 + z^2}{1 - z^2} \right|.$$

This can be recognized as Scherk's surface, as presented in Section 9.4, with the scale factor $\sqrt{2}/\pi$.

Finally, instead of lifting harmonic mappings to construct minimal surfaces, the process can be reversed. It is sometimes fruitful to project specific minimal surfaces to arrive at new families of harmonic mappings. For instance, one can start with Scherk's "saddle-tower" minimal surface

$$\sin t = \sinh u \sinh v,$$

known variously as *Scherk's second surface* or *Scherk's fifth surface*, whose graph is shown in Figure 10.1. Its Weierstrass–Enneper representation is known to be given by $p(z) = 1/(1 - z^4)$ and $q(z) = z$, suggesting a close relation with Scherk's first surface. (In fact, the two surfaces are conjugate, as are the catenoid and helicoid.) Carrying out the integrations by partial fractions and projecting the surface to the base plane, one arrives at the harmonic function

$$w = u + iv = f(z) = -\frac{1}{2} \sum_{k=1}^{4} i^k \log |z - i^k|,$$

which maps the unit disk onto a starlike region with infinite spires along the coordinate axes as depicted in Figure 10.2. The harmonic mapping just

Figure 10.1. Scherk's saddle-tower surface.

constructed can be generalized to

$$w = F_n(z) = -\frac{2}{n} \sum_{k=1}^{n} \alpha^k \log |z - \alpha^k|, \qquad n = 3, 4, 5, \ldots,$$

where $\alpha = e^{2\pi i/n}$ is a primitive nth root of unity. By appeal to the argument principle for harmonic functions (see Section 1.3), it can be shown that F_n maps the disk *univalently* onto a rotationally symmetric starlike region with

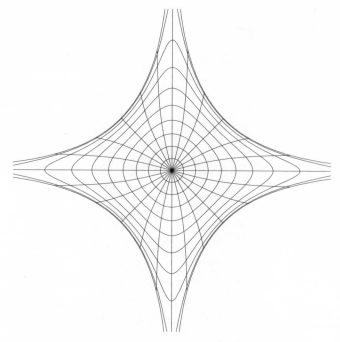

Figure 10.2. Harmonic mapping for Scherk's saddle tower.

infinite spires in the directions of the nth roots of unity $1, \alpha, \alpha^2, \ldots, \alpha^{n-1}$. The dilatation of F_n is found to be $\omega_n(z) = z^{n-2}$. Consequently, in view of the theorem just proved, F_n has a canonical lifting to a minimal surface if and only if n is even. For $n = 6, 8, 10, \ldots$, these surfaces constitute generalizations of Scherk's saddle-tower surface. Enneper's surface and Scherk's first surface can be generalized in an analogous way. Further details are contained in a paper by Duren and Thygerson [1].

10.3. Heinz's Lemma and Bounds on Curvature

According to the theorem of the previous section, the projections of minimal graphs in isothermal parameters are precisely the harmonic mappings whose dilatations are squares of meromorphic functions. If S is a minimal surface lying over a simply connected domain Ω in the uv–plane, expressed in isothermal parameters (x, y), its projection onto the base plane may be interpreted as a harmonic mapping $w = f(z)$, where $w = u + iv$ and $z = x + iy$. After suitable adjustment of parameters, it may be assumed that f is a sense-preserving harmonic mapping of the unit disk $\mathbb{D}$ onto Ω, with $f(0) = w_0$

for some preassigned point w_0 in Ω. Let $f = h + \overline{g}$ be the canonical decomposition. Then the dilatation $\omega = g'/h'$ of f is an analytic function with $|\omega(z)| < 1$ in $\mathbb{D}$ and with the further property that $\omega = q^2$ for some function q analytic in $\mathbb{D}$.

The minimal surface S over Ω has the isothermal representation

$$u = \mathrm{Re}\{f(z)\} = \mathrm{Re}\left\{ \int_0^z \varphi_1(\zeta)\, d\zeta \right\},$$

$$v = \mathrm{Im}\{f(z)\} = \mathrm{Re}\left\{ \int_0^z \varphi_2(\zeta)\, d\zeta \right\},$$

$$t = \mathrm{Re}\left\{ \int_0^z \varphi_3(\zeta)\, d\zeta \right\}$$

for $z \in \mathbb{D}$, with

$$\varphi_1 = h' + g' = p(1 + q^2), \qquad \varphi_2 = -i(h' - g') = -ip(1 - q^2),$$

and $\varphi_3 = -2ipq$, where p and q are the Weierstrass–Enneper parameters. Thus $\varphi_3{}^2 = -4\omega h'^2$ and $h' = p$.

The first fundamental form of S is $ds^2 = \lambda^2 |dz|^2$, where

$$\lambda^2 = \frac{1}{2} \sum_{k=1}^3 |\varphi_k|^2.$$

A direct calculation shows that

$$\lambda^2 = |h'|^2 + |g'|^2 + 2|g'h'| = (|h'| + |g'|)^2,$$

so that $\lambda = |h'| + |g'| = |p|(1 + |q|^2)$.

This simple expression allows us to calculate the Gauss curvature of S in terms of the underlying harmonic mapping. Note that $p(z) = h'(z) \neq 0$ in $\mathbb{D}$ since f is sense-preserving. As found in Section 10.1, the general formula for Gauss curvature is $K = -\lambda^{-2}\Delta(\log \lambda)$, where Δ denotes the Laplacian. Recalling that $\Delta = 4\partial^2/\partial\overline{z}\partial z$, we first find

$$\frac{\partial}{\partial z}(\log |p|) = \frac{1}{2}\frac{\partial}{\partial z}\{\log p + \log \overline{p}\} = \frac{p'}{2p},$$

so that $\Delta(\log |p|) = 0$. Next,

$$\frac{\partial}{\partial z}\{\log(1 + |q|^2)\} = \frac{\partial}{\partial z}\{\log(1 + q\overline{q})\} = \frac{q'\overline{q}}{1 + |q|^2},$$

and so

$$\Delta\{\log(1 + |q|^2)\} = \frac{4|q'|^2}{(1 + |q|^2)^2}.$$

Therefore, in terms of the Weierstrass–Enneper parameters, the Gauss curvature is found to be

$$K = -\frac{4|q'|^2}{|p|^2(1 + |q|^2)^4}.$$

Since the underlying harmonic mapping f has dilatation $\omega = g'/h' = q^2$, and $h' = p$, an equivalent expression is

$$K = -\frac{|\omega'|^2}{|h'g'|(1 + |\omega|)^4}.$$

The formula allows the intervention of analytic function theory to estimate Gauss curvature. Since $|q(z)| < 1$, the Schwarz–Pick lemma gives

$$\frac{|q'(z)|}{1 - |q(z)|^2} \leq \frac{1}{1 - |z|^2}, \qquad z \in \mathbb{D}.$$

Therefore, at the point of the surface that lies above $w_0 = f(0)$, we are led to the estimate

$$|K| \leq \frac{4(1 - |q(0)|^2)^2}{|p(0)|^2(1 + |q(0)|^2)^4} = \frac{4(1 - |\omega(0)|)^2}{(|h'(0)| + |g'(0)|)^2(1 + |\omega(0)|)^2}$$

$$\leq \frac{4}{(|h'(0)| + |g'(0)|)^2} \leq \frac{4}{|h'(0)|^2 + |g'(0)|^2}.$$

Let us now specialize the problem by supposing that $\Omega = \mathbb{D}$ and $w_0 = 0$. In other words, S is a minimal graph above the unit disk, and K is the Gauss curvature at the point on the surface above the origin. The projection of S is then a harmonic mapping of $\mathbb{D}$ onto $\mathbb{D}$ with $f(0) = 0$, so the sharp form of Heinz' lemma (see Section 4.4) provides the estimate

$$|h'(0)|^2 + |g'(0)|^2 \geq \frac{27}{4\pi^2}.$$

In view of the preceding inequality, this gives the bound

$$|K| \leq \frac{16\pi^2}{27} = 5.848.\dots$$

However, the bound is not sharp. The estimate used the sharp form of Heinz' lemma, but equality occurs there for a function f that maps the disk onto an inscribed equilateral triangle, with dilatation $\omega = \overline{f_{\bar{z}}}/f_z$ given by $\omega(z) = z$. This dilatation is not the square of an analytic function, so in view of the theorem in Section 10.2, the mapping f cannot lift to a minimal surface. Strictly speaking, this mapping does not qualify because its range is not the full disk. However, it maps the disk onto itself in the weak sense, with $f(\mathbb{D}) \subset \mathbb{D}$ and radial limits $f(e^{i\theta}) \in \mathbb{T}$ almost everywhere; and f can be approximated locally uniformly in $\mathbb{D}$ by harmonic mappings of the disk onto itself. These approximating mappings show that the bound $27/4\pi^2$ in Heinz' lemma is sharp, but again they cannot lift to minimal surfaces.

In further demonstration that the bound $16\pi^2/27$ for curvature is not sharp, Richard Hall [4] has given a very small numerical improvement.

For the sharp estimate of curvature, a constrained form of Heinz' lemma is required. Specifically, it is required to find the sharp lower bound of $|h'(0)|^2 + |g'(0)|^2$ among all harmonic self-mappings of the disk with $f(0) = 0$ and dilatation $\omega = q^2$ for some analytic function q. It is reasonable to conjecture that the "extremal function" is now the canonical mapping of the disk onto an inscribed square (see Section 4.2), with dilatation $\omega(z) = z^2$. This mapping has $a_1 = h'(0) = 2\sqrt{2}/\pi$ and $b_1 = g'(0) = 0$, so the constrained form of Heinz' lemma can be expected to run as follows.

Conjecture 1. Let f be a harmonic mapping of $\mathbb{D}$ onto $\mathbb{D}$ with $f(0) = 0$, whose dilatation $\omega = \overline{f_{\bar{z}}}/f_z$ is the square of an analytic function. Then

$$|f_z(0)|^2 + |f_{\bar{z}}(0)|^2 > \frac{8}{\pi^2}$$

and the bound is sharp.

The canonical mapping onto the square lifts to Scherk's first surface, as was observed in Section 9.4. The Weierstrass–Enneper parameters for Scherk's surface, adjusted to lie above a square inscribed in the unit circle, are

$$p(z) = \frac{2\sqrt{2}}{\pi(1 - z^4)} \qquad \text{and} \qquad q(z) = iz.$$

Thus, at the point above the origin the surface has curvature

$$K = -\frac{4|q'(0)|^2}{|p(0)|^2(1 + |q(0)|^2)^4} = -\frac{\pi^2}{2}.$$

This suggests the following sharp form of the curvature estimate.

Conjecture 2. For any minimal graph lying above the entire unit disk, the Gauss curvature at the point above the origin satisfies the sharp inequality $|K| < \frac{\pi^2}{2} = 4.934\ldots$.

If Conjecture 1 were known to be true, then Conjecture 2 would follow by the same argument used to obtain the estimate $|K| \leq \frac{16\pi^2}{27}$. In fact, the two conjectures are equivalent. Finn and Osserman [1] proved Conjecture 2 under the additional assumption that the minimal surface has a horizontal tangent plane at the point above the origin.

A nonparametric portion of Enneper's surface, optimally normalized so that it just covers the whole unit disk, has parameters $p(z) = \frac{3}{2}$ and $q(z) = iz$ (*cf.* Section 9.4), so its curvature at the point above the origin is $K = -32/9 = -3.555\ldots$.

Finally, the upper bound for curvature has an interesting consequence. We have shown that if a minimal graph lies above the entire unit disk, then its Gauss curvature at the point above the origin is bounded by the absolute constant $C = 16\pi^2/27$. From this it follows more generally that whenever a minimal graph covers a full disk of radius R, then its curvature at the point above the center of that disk satisfies $|K| \leq \frac{C}{R^2}$. Consequently, if a minimal graph actually lies above the entire plane, its Gauss curvature at every point is $K = 0$. Since the mean curvature of a minimal surface also vanishes, it follows that any such minimal graph must have both principal curvatures equal to zero at every point. This proves a classical theorem of S. Bernstein.

Bernstein's Theorem. *A minimal graph that lies above the entire plane must itself be a plane.*

10.4. Sharp Bounds on Curvature

In the preceding section we found an upper bound for the magnitude of Gauss curvature of a minimal surface that lies above the unit disk, computed at the point of the surface above the center of the disk. Although the sharp bound is not known, the corresponding sharp bounds can be found for other regions such as a half-plane or an infinite strip or the whole plane with a linear slit. These results are due to Hengartner and Schober [7]. With the observation that each of the domains is convex in the horizontal direction, the main idea is to apply the shear construction to relate the underlying harmonic mappings to conformal mappings. The idea works in each case because the conformal "preshear" of the "extremal" harmonic mapping turns out to map the disk *onto* the given domain. This is no longer true when the disk is the target domain. Here we shall treat the problem only for an infinite strip, since the discussion for the other two domains is quite similar.

Let $\Omega = \{w \in \mathbb{C} : |\text{Im}\{w\}| < \frac{\pi}{4}\}$ denote an infinite strip of width $\frac{\pi}{2}$ and let $\alpha = a + ib$ be a specified point in Ω. For an arbitrary minimal surface over Ω, our problem is to find the sharp upper bound for $|K|$, where K is the Gauss curvature of the minimal surface at the point above α. It is intuitively clear that the bound will depend only on $|b|$.

Now we introduce the Weierstrass–Enneper representation of the surface (Section 9.3, Theorem 2), where the parameters p and q are analytic in the unit disk $\mathbb{D}$, with $|q(z)| < 1$ there. This parametric representation of the surface projects to a sense-preserving harmonic mapping $f = h + \overline{g}$ of $\mathbb{D}$ onto Ω, which may be normalized so that $g(0) = 0$ and $f(0) = \alpha$. Recall that f has dilatation $\omega = g'/h' = q^2$ and that $h' = p$, while the Gauss curvature at the point above α satisfies

$$|K| \leq \frac{4(1 - |q(0)|^2)^2}{|h'(0)|^2(1 + |q(0)|^2)^4},$$

as shown in the preceding section.

We now invoke the shear construction (Section 3.4, Theorem 1). Since $f = h + \overline{g}$ is a harmonic mapping of $\mathbb{D}$ onto a domain Ω that is convex in the horizontal direction (CHD), the associated analytic function $\varphi = h - g$ maps $\mathbb{D}$ conformally into Ω and is also CHD. Note that $\varphi(0) = \alpha$ and that

$$\varphi' = h' - g' = (1 - q^2)h',$$

so that the curvature estimate takes the form

$$|K| \leq \frac{4(1 - |q(0)|^2)^2|1 - q(0)^2|^2}{(1 + |q(0)|^2)^4|\varphi'(0)|^2} \leq \frac{4}{|\varphi'(0)|^2}.$$

The next step is to verify that the function

$$\psi(z) = \alpha + \frac{1}{2}\log\frac{1 + \zeta z}{1 - z}, \qquad \zeta = e^{-4ib},$$

maps $\mathbb{D}$ conformally onto Ω with $\psi(0) = \alpha$. (Note that $|b| < \pi/4$, so that $\zeta \neq -1$.) Since φ maps $\mathbb{D}$ *into* Ω and has all but two of its boundary values on $\partial\Omega$, we see that φ actually maps $\mathbb{D}$ *onto* Ω. Therefore, $\varphi(z) = \psi(\gamma z)$ for some complex constant γ of unit modulus. But since a rotation of the disk simply amounts to a reparametrization of the surface, there is no loss of generality in taking $\gamma = 1$, so that $\varphi = \psi$. Now an easy calculation gives

$$\varphi'(0) = \frac{1}{2}(1 + \zeta) = e^{-2ib}\cos 2b,$$

and it follows that

$$|K| \leq 4\sec^2(2b).$$

To investigate the possibility of equality, we first observe that equality occurs in our estimates for curvature if and only if the underlying harmonic mapping f has dilatation $\omega(z) = \lambda z^2$ for some constant λ with $|\lambda| = 1$. The question is now whether such a function f will lift to a minimal surface lying over the entire strip Ω, or equivalently whether the harmonic mapping f, obtained from $\varphi = \psi$ by shearing with dilatation λz^2, will actually map $\mathbb{D}$ *onto* Ω. But this function $f = h + \overline{g}$ satisfies $h' - g' = \varphi'$ and $\omega h' - g' = 0$, so that

$$h'(z) = \frac{\varphi'(z)}{1 - \omega(z)} = \frac{\zeta + 1}{2(1 - \lambda z^2)(1 + \zeta z)(1 - z)}.$$

Thus the function h with $h(0) = \alpha$ can be found by integration, and then $g = h - \varphi$.

The simplest case is where $\alpha = a$ is real, so that $b = 0$ and $\zeta = 1$. If we take $\lambda = 1$, we have

$$h'(z) = \frac{1}{(1 - z^2)^2},$$

and integration gives

$$h(z) = \frac{1}{4} \log \frac{1 + z}{1 - z} + \frac{1}{2} \frac{z}{1 - z^2} + a,$$

$$g(z) = -\frac{1}{4} \log \frac{1 + z}{1 - z} + \frac{1}{2} \frac{z}{1 - z^2}.$$

Thus $f = u + iv$, where

$$u(z) = \mathrm{Re}\left\{ \frac{z}{1 - z^2} \right\}, \qquad v(z) = \frac{1}{2} \arg \left\{ \frac{1 + z}{1 - z} \right\}.$$

By the Weierstrass–Enneper formulas, the height of the minimal surface above $f(z)$ is

$$t = 2 \, \mathrm{Im}\left\{ \int \frac{z}{(1 - z^2)^2} \, dz \right\} = \mathrm{Im}\left\{ \frac{1}{1 - z^2} \right\}.$$

To see that $f(\mathbb{D}) = \Omega$, we note that $z \mapsto \frac{1+z}{1-z}$ maps the unit disk onto the right half-plane, and we make the following change of coordinates:

$$Re^{i\theta} = \frac{1 + z}{1 - z}, \qquad R > 0, \quad |\theta| < \frac{\pi}{2}.$$

Then

$$u - a = \frac{1}{4}\left(R - \frac{1}{R} \right) \cos \theta, \qquad v = \frac{\theta}{2}, \qquad t = \frac{1}{4}\left(R - \frac{1}{R} \right) \sin \theta.$$

This shows that f maps the disk onto Ω, and the minimal surface is a *helicoid*. Thus the upper bound is attained and the estimate $|K| \leq 4$ is sharp when α is real.

If α is not real, it turns out that the shear construction with dilatation λz^2 always produces, for any λ with $|\lambda| = 1$, a *bounded* function f, mapping $\mathbb{D}$ onto a proper subdomain of Ω, so that the resulting minimal surface does *not* lie over the whole strip. However, the device of replacing λ by $r\lambda$, where $0 < r < 1$, generates a minimal surface that does lie over all of Ω, whose Gauss curvature at the point above α approaches $-4 \sec^2(2b)$ as r tends to 1. Thus, the bound is sharp for nonreal points α as well, but it is never actually attained by a minimal surface over the whole strip Ω.

Full details can be found in the paper by Hengartner and Schober [7]. The final result may be summarized as follows.

Theorem. *Let S be a nonparametric minimal surface over the infinite strip Ω defined by $|\text{Im}\{w\}| < \frac{\pi}{4}$ and let $\alpha = a + ib$ be an arbitrary point in Ω. Then the Gauss curvature K of S at the point above α satisfies the sharp inequalities $|K| \leq 4$ if $b = 0$, and $|K| < 4 \sec^2(2b)$ if $b \neq 0$.*

10.5. Schwarzian Derivatives

The *Schwarzian derivative* of a locally univalent analytic function f is defined by

$$S(f) = (f''/f')' - \frac{1}{2}(f''/f')^2.$$

The key property is its invariance under postcomposition with Möbius transformations: $S(T \circ f) = S(f)$ for all Möbius (or linear fractional) transformations

$$T(z) = \frac{az + b}{cz + d}, \qquad ad - bc \neq 0.$$

Although this property is easily verified, it is natural to ask how it might have been discovered. Here is the derivation essentially given by H. A. Schwarz in 1873. Suppose $g = T \circ f$ for some Möbius transformation T, so that $(cf + d)g = af + b$. Three successive differentiations produce the system of linear equations

$$c(fg)' + dg' - af' = 0$$
$$c(fg)'' + dg'' - af'' = 0$$
$$c(fg)''' + dg''' - af''' = 0,$$

with nontrivial solution $(c, d, -a)$, so the determinant of coefficients vanishes identically. When the determinant is expanded, the equation reduces to

$$3g'^2 f''^2 + 2g'g'''f'^2 = 3f'^2 g''^2 + 2f'f'''g'^2.$$

Dividing both sides by $2f'^2 g'^2$, we conclude that

$$g'''/g' - \frac{3}{2}(g''/g')^2 = f'''/f' - \frac{3}{2}(f''/f')^2,$$

which says that $S(g) = S(f)$.

The invariance property $S(T \circ f) = S(f)$ is a special case of the composition formula

$$S(g \circ f) = (S(g) \circ f)f'^2 + S(f)$$

for arbitrary analytic functions f and g, since $S(T) = 0$ for Möbius transformations. For an arbitrary analytic function φ, the set of functions f with Schwarzian $S(f) = 2\varphi$ can be described by $f = w_1/w_2$, where w_1 and w_2 are linearly independent solutions of the *linear* differential equation $w'' + \varphi w = 0$. Two consequences are as follows:

(i) If $S(f) = 0$, then f is a Möbius transformation.
(ii) If $S(g) = S(f)$, then $g = T \circ f$ for some Möbius transformation T.

In 1949, Z. Nehari [1] exploited the connection with linear differential equations to obtain important criteria for global univalence expressed in terms of the Schwarzian derivative. For instance, if f is analytic and locally univalent in $\mathbb{D}$ and if its Schwarzian derivative satisfies

$$|S(f)(z)| \le \frac{2}{(1 - |z|^2)^2}, \qquad z \in \mathbb{D},$$

then f is univalent in $\mathbb{D}$. Nehari also showed that the uniform bound $|S(f)(z)| \le \pi^2/2$ implies the univalence of f in $\mathbb{D}$. Other univalence criteria of similar type have been discovered. Proofs and further discussion may be found in Duren [2].

It is interesting to ask whether these univalence criteria can be generalized to harmonic mappings. The first problem, however, is to find a suitable definition of Schwarzian derivative for locally univalent harmonic functions. A natural definition has been proposed by Chuaqui, Duren, and Osgood [1], exploiting the differential geometry of the associated minimal surface. If $f = h + \overline{g}$ is harmonic, locally univalent, and sense-preserving, and if its dilatation $\omega = g'/h'$ has the form $\omega = q^2$ for some analytic function q, then it can be lifted locally to a minimal surface with conformal metric $ds = \lambda |dz|$, where

$$\lambda = |h'| + |g'| = |p|(1 + |q|^2)$$

in terms of the Weierstrass–Enneper functions p and q. The Schwarzian derivative of f is defined by the formula

$$S(f) = 2\{(\log \lambda)_{zz} - ((\log \lambda)_z)^2\}.$$

If f is *analytic*, then $\lambda = |f'|$, so that

$$\log \lambda = \frac{1}{2}(\log f' + \log \overline{f'}).$$

Thus $(\log \lambda)_z = \frac{1}{2} f''/f'$, so the generalized Schwarzian is

$$S(f) = 2\{(\log \lambda)_{zz} - ((\log \lambda)_z)^2\} = (f''/f')' - \frac{1}{2}(f''/f')^2,$$

which is in agreement with the classical formula.

In general, we can write $\lambda = |h'|(1 + q\overline{q})$, so that

$$(\log \lambda)_z = \frac{1}{2} \frac{h''}{h'} + \frac{q'\overline{q}}{1 + |q|^2},$$

and the Schwarzian is

$$S(f) = S(h) + \frac{2\overline{q}}{1 + |q|^2}\left(q'' - \frac{q'h''}{h'}\right) - 4\left(\frac{q'\overline{q}}{1 + |q|^2}\right)^2.$$

Note that $S(f) = S(h)$ if $\omega = q^2$ is constant. But we know (*cf.* Section 7.1) that if $\omega(z) \equiv \alpha$, where α is a complex constant with $|\alpha| < 1$, then $f = h + \alpha \overline{h}$ for some analytic function h. Thus $S(h + \alpha \overline{h}) = S(h)$, a fact that is easy to verify by direct calculation.

If φ is an analytic function for which the composition $f \circ \varphi$ is defined, then $f \circ \varphi$ is again a locally univalent harmonic function with dilatation $q \circ \varphi^2$, and

$$\lambda_{f \circ \varphi} = (\lambda_f \circ \varphi)|\varphi'|.$$

A calculation then gives

$$S(f \circ \varphi) = (S(f) \circ \varphi)\varphi'^2 + S(\varphi),$$

which is a generalization of the classical transformation formula for Schwarzians of analytic functions under composition.

Recall now (*cf.* Section 10.3) that the Gauss curvature K of the minimal surface associated with a harmonic mapping $f = h + \overline{g}$, with dilatation $\omega = g'/h' = q^2$, is given by the formula

$$K = -\frac{4|q'|^2}{|p|^2(1 + |q|^2)^4},$$

where $p = h'$ and q are the Weierstrass–Enneper functions. We now show that the Schwarzian $S(f)$ is analytic only for harmonic functions of the form $f = h + \alpha \overline{h}$ with h analytic and $|\alpha| < 1$ or, equivalently, when the associated minimal surface is a plane. Specifically, we shall prove the following theorem.

Theorem 1. *For a locally univalent sense-preserving harmonic function f with dilatation $\omega = q^2$, the following are equivalent:*

- *(i) $S(f)$ is analytic.*
- *(ii) The curvature K of the minimal surface locally associated with f is constant.*
- *(iii) $K \equiv 0$ so that the corresponding minimal surface is a plane.*
- *(iv) The dilatation of f is constant.*
- *(v) $f = h + \alpha \overline{h}$ for some analytic locally univalent function h and for some complex constant α with $|\alpha| < 1$.*

Proof. (i) $\Longrightarrow$ (ii). The curvature of the minimal surface associated with f is

$$K = -\frac{1}{\lambda^2} \, \Delta(\log \lambda) = -\frac{4(\log \lambda)_{z\overline{z}}}{\lambda^2} \, .$$

A simple calculation yields

$$-\frac{1}{4} K_z = \frac{1}{\lambda^2} \left[(\log \lambda)_{zz} - ((\log \lambda)_z)^2 \right]_{\overline{z}} = \frac{1}{2\lambda^2} [S(f)]_{\overline{z}} = 0$$

if $S(f)$ is analytic. Thus K is constant.

(ii) $\Longrightarrow$ (iii). Referring to the formula for curvature in terms of the Weierstrass–Enneper parameters p and q, and passing to logarithms, we see that if K is constant then

$$\log(1 + |q|^2) = \frac{1}{2} \, \log |q'/p| + c$$

for some constant c. Thus $\log(1 + |q|^2)$ is a harmonic function. But a calculation gives

$$\left[\log(1 + |q|^2)\right]_{z\overline{z}} = \left[\frac{q' \overline{q}}{1 + |q|^2} \right]_{\overline{z}} = \frac{|q'|^2}{(1 + |q|^2)^2} \, ,$$

so $\log(1 + |q|^2)$ is harmonic if and only if $q' = 0$. But then the formula for curvature shows that $K = 0$.

(iii) $\implies$ (iv). If $K = 0$, then $q' = 0$, so q is constant and the dilatation $\omega = q^2$ is constant.

(iv) $\implies$ (v). This was already noted above, but here are further details. If a harmonic mapping $f = h + \overline{g}$ has constant dilatation, then $g' = \alpha h'$ for some constant α with $|\alpha| < 1$. Integration gives $g = \alpha h + \beta$ for some constant β. But since $|\alpha| \neq 1$, a bit of linear algebra shows that the additive constant β can be absorbed into h and we can write, with slight change of notation, $f = h + \alpha \overline{h}$ for some analytic locally univalent function h.

(v) $\implies$ (i). As observed earlier, $S(h + \alpha \overline{h}) = S(h)$. ∎

Recall now that the *analytic* functions with vanishing Schwarzian derivatives are precisely the Möbius transformations. With appeal to Theorem 1, we can now obtain a corresponding result for harmonic mappings.

Theorem 2. *A sense-preserving harmonic function f has vanishing Schwarzian derivative $S(f) = 0$ if and only if it has the form $f = h + \alpha \overline{h}$ for some Möbius transformation h and some complex constant α with $|\alpha| < 1$.*

Proof. If $f = h + \alpha \overline{h}$ for a Möbius transformation h, then $S(f) = S(h) = 0$. Conversely, suppose that a harmonic mapping $f = h + \overline{g}$ has Schwarzian derivative $S(f) = 0$. Then by Theorem 1 we can conclude that f has constant dilatation $\omega = g'/h'$, so $|g'| = c|h'|$ for some constant $c \geq 0$. It follows that

$$\lambda = |h'| + |g'| = (1 + c)|h'|,$$

so that $0 = S(f) = S(h)$ and h is a Möbius transformation. Also, since $\omega = \alpha$ for some constant α with $|\alpha| < 1$, we see that $g = \alpha h + \beta$ for some constant β. Again the additive constant β can be absorbed into the Möbius transformation h, and so with change of notation we can write $f = h + \alpha \overline{h}$ for some Möbius transformation h. ∎

The theorem shows that a sense-preserving harmonic function with $S(f) = 0$ is globally univalent and extends to a harmonic mapping of $\mathbb{C}$ onto itself. We define a *harmonic Möbius transformation* to be a harmonic mapping of the form $f = h + \alpha \overline{h}$, where h is a (classical) Möbius transformation and α is a complex constant with $|\alpha| < 1$. Theorem 2 says that these are precisely the harmonic mappings with $S(f) = 0$. Since a harmonic Möbius transformation is the composition of a Möbius transformation with an affine mapping, we can see that a harmonic Möbius transformation maps circles to ellipses. The basic composition formula shows that $S(f \circ \varphi) = S(\varphi)$ if φ is analytic and f is a harmonic Möbius transformation.

The next problem is to describe the relation between two harmonic mappings that have the same Schwarzian derivative. One form of the solution is given by the following theorem, where curvatures of the associated conformal metrics play an essential role.

Theorem 3. *Let $f = h + \overline{g}$ and $F = H + \overline{G}$ be sense-preserving harmonic functions defined on a common domain $\Omega \subset \mathbb{C}$. If $S(f) = S(F)$, then*

    (a) *The curvatures of the associated conformal metrics are equal: $K(\lambda_f) = K(\lambda_F)$.*

    (b) *If the curvatures are not constant, then the metrics are homothetic; that is, $\lambda_f = c \, \lambda_F$ for some constant $c > 0$.*

    (c) *If the curvatures are constant, then both are zero, and $f = h + \alpha \overline{h}$, $F = H + \beta \overline{H}$, and $H = T(h)$ for some analytic univalent functions h and H, some complex constants α and β with $|\alpha| < 1$ and $|\beta| < 1$, and some analytic Möbius transformation T.*

Conversely, if either (b) or (c) holds, then the curvatures are equal and $S(f) = S(F)$.

The proof requires some further geometric background and will not be pursued here. The paper by Chuaqui, Duren, and Osgood [1] gives the proof and contains further information.

In order to display a few explicit examples, we shall now calculate the Schwarzian derivatives of particular harmonic mappings that have been discussed elsewhere in this book.

First consider the harmonic mapping $f(z) = z + \frac{1}{3}\overline{z}^3$, which has dilatation $\omega(z) = z^2$ and maps the unit disk onto the domain inside a hypocycloid of four cusps inscribed in the circle $|w| = \frac{4}{3}$. Here $\lambda = |h'| + |g'| = 1 + |z|^2$, so the Schwarzian derivative is

$$S(f) = -\frac{4\,\overline{z}^2}{(1 + |z|^2)^2} \, .$$

Next consider the harmonic mapping

$$f(z) = \log\left|\frac{1 + z}{1 - z}\right| - \overline{z},$$

which results from horizontal shearing of the conformal mapping $\varphi(z) = z$ with dilatation $\omega(z) = z^2$ (see Section 3.4). Here $f = h + \overline{g}$, where

$$h(z) = \frac{1}{2} \log \frac{1+z}{1-z}, \qquad g(z) = h(z) - z.$$

Thus

$$\lambda = |h'| + |g'| = \frac{1 + |z|^2}{|1 - z^2|},$$

and a simple calculation gives the Schwarzian derivative

$$S(f) = 2\left(\frac{1}{(1-z^2)^2} - \frac{2\overline{z}^2}{(1+|z|^2)^2} - \frac{2|z|^2}{(1+|z|^2)(1-z^2)} \right).$$

As shown in Section 3.4, the general harmonic shear of a conformal mapping φ convex in the horizontal direction, with dilatation $\omega = q^2$, has the form $f = h + \overline{g}$, where $h - g = \varphi$ and $g' = q^2 h'$. Solving the pair of linear equations, one finds $h' = \varphi'/(1 - q^2)$. A calculation then yields the formula

$$S(f) = S(\varphi) + \frac{2(q'^2 + (1-q^2)qq'')}{(1-q^2)^2} - \frac{2qq'}{1-q^2} \frac{\varphi''}{\varphi'}$$

$$+ \frac{2\overline{q}}{1+|q|^2} \left\{ q'' - q'\left(\frac{\varphi''}{\varphi'} + \frac{2qq'}{1-q^2} \right) \right\} - 4\left(\frac{q'\overline{q}}{1+|q|^2} \right)^2.$$

If φ is the Koebe function $k(z) = z/(1-z)^2$ and $q(z) = z$, the formula reduces to

$$S(f) = -4\left(\frac{1}{(1-z)^2} + \frac{\overline{z}}{1+|z|^2} \right)^2.$$

It seems likely that Schwarzian derivatives will be useful in the further study of harmonic mappings, particularly for questions of global univalence.

Appendix
Extremal Length

Extremal length is a conformal invariant that has gained broad acceptance as a tool in geometric function theory. Its origins can be traced to work of Herbert Grötzsch as early as 1928, but the modern formulation was introduced by Ahlfors and Beurling around 1950. This appendix gives a brief introduction to extremal length, by way of background for the application in Section 6.2 to a covering theorem for harmonic mappings. More extensive treatments can be found in the books by Ahlfors [2] and Fuchs [1].

Let Ω be a domain in the complex plane and let Γ be a family of locally rectifiable arcs γ in Ω. A *metric* is a Borel measurable function $\rho(z) \geq 0$ on Ω. The *ρ-length* of an arc $\gamma \in \Gamma$ is

$$L(\gamma) = \int_\gamma \rho(z)\,|dz|.$$

A metric ρ is said to be *admissible* for the curve family Γ if $L(\gamma) \geq 1$ for every $\gamma \in \Gamma$. The *extremal length* of Γ is the quantity $\lambda(\Gamma)$ defined by

$$\frac{1}{\lambda(\Gamma)} = \inf \iint_\Omega \rho(z)^2\,dx\,dy, \qquad z = x + iy,$$

where the infimum is taken over all admissible metrics ρ. Thus, $0 \leq \lambda(\Gamma) \leq \infty$. Note that the domain Ω does not play an essential role, since the metrics can be taken to vanish off the support of the arcs in Γ.

The *comparison principle* says that if $\tilde{\Gamma}$ is another curve family in Ω with $\tilde{\Gamma} \subset \Gamma$ or, more generally, if every arc $\tilde{\gamma} \in \tilde{\Gamma}$ has a subarc $\gamma \in \Gamma$, then $\lambda(\Gamma) \leq \lambda(\tilde{\Gamma})$. To see this, it is enough to observe that any metric admissible for Γ must also be admissible for $\tilde{\Gamma}$. Thus, the infimum $1/\lambda(\tilde{\Gamma})$ of the area integral taken over all metrics ρ admissible for $\tilde{\Gamma}$ is less than or equal to the infimum $1/\lambda(\Gamma)$ taken over the metrics ρ admissible for Γ. This shows that $\lambda(\Gamma) \leq \lambda(\tilde{\Gamma})$.

The *composition laws* give more delicate inequalities. Suppose that Ω_1 and Ω_2 are disjoint domains and that Γ_1 and Γ_2 are families of arcs in Ω_1 and

Ω_2, respectively. Then the composition laws are as follows:

(i) If every arc $\gamma \in \Gamma$ has a subarc $\gamma_1 \in \Gamma_1$ and a subarc $\gamma_2 \in \Gamma_2$, then
$$\lambda(\Gamma_1) + \lambda(\Gamma_2) \le \lambda(\Gamma).$$

(ii) If every arc $\gamma_1 \in \Gamma_1$ and every arc $\gamma_2 \in \Gamma_2$ has a subarc $\gamma \in \Gamma$, then
$$\frac{1}{\lambda(\Gamma_1)} + \frac{1}{\lambda(\Gamma_2)} \le \frac{1}{\lambda(\Gamma)}.$$

The proofs are not difficult, but they will be omitted here.

The most important property of extremal length is its *conformal invariance*. Suppose the domain Ω is mapped conformally onto a domain $\tilde{\Omega}$ by some function φ carrying the arcs $\gamma \in \Gamma$ to arcs $\tilde{\gamma} = \varphi(\gamma)$ that comprise a family $\tilde{\Gamma}$ in $\tilde{\Omega}$. Let $z = \psi(w)$ be the inverse mapping of $\tilde{\Omega}$ onto Ω. If a metric $\rho(z)$ is admissible for Γ, then

$$1 \le \int_\gamma \rho(z)\,|dz| = \int_{\tilde{\gamma}} \rho(\psi(w))\,|\psi'(w)|\,|dw|$$

for every $\gamma \in \Gamma$ and, hence, for every $\tilde{\gamma} \in \tilde{\Gamma}$, so the metric $\tilde{\rho}(w) = \rho(\psi(w))\,|\psi'(w)|$ is admissible for $\tilde{\Gamma}$. On the other hand,

$$\iint_\Omega \rho(z)^2\,dx\,dy = \iint_{\tilde{\Omega}} \rho(\psi(w))^2|\psi'(w)|^2\,du\,dv = \iint_\Omega \tilde{\rho}(w)^2\,du\,dv,$$

where $w = u + iv$, since $|\psi'(w)|^2$ is the Jacobian of the mapping $z = \psi(w)$. Taking the infimum over all metrics ρ admissible for Γ, we deduce that $1/\lambda(\Gamma) \ge 1/\lambda(\tilde{\Gamma})$ because the metrics $\tilde{\rho}$ of the above form may conceivably form a proper subset of the full set of metrics admissible for $\tilde{\Gamma}$. Thus, $\lambda(\Gamma) \le \lambda(\tilde{\Gamma})$. In fact, every metric $\tilde{\rho}$ admissible for $\tilde{\Gamma}$ has the above form for some ρ admissible for Γ, but we need not bother to confirm that since we can reverse the roles of Γ and $\tilde{\Gamma}$ and conclude in the same manner that $\lambda(\tilde{\Gamma}) \le \lambda(\Gamma)$. Consequently, $\lambda(\Gamma) = \lambda(\tilde{\Gamma})$, which proves the conformal invariance of extremal length.

To give explicit examples, let $0 < a < b < \infty$ and let Ω be the annulus defined by $a < |z| < b$. First take Γ to be the family of curves that connect the two boundary components. Since each radial segment $\{z = re^{i\theta} : a < r < b\}$ belongs to the family Γ, we see that $\int_a^b \rho(re^{i\theta})\,dr \ge 1$ for each $\theta \in [0, 2\pi]$ if the metric ρ is admissible for Γ. Applying the Schwarz inequality, we infer that

$$1 \le \left\{ \int_a^b \rho(re^{i\theta})\,dr \right\}^2 \le \int_a^b \rho(re^{i\theta})^2\,r\,dr \int_a^b \frac{1}{r}\,dr,$$

or

$$1 \le \log\frac{b}{a} \int_a^b \rho(re^{i\theta})^2\,r\,dr.$$

Integration with respect to θ therefore gives

$$\frac{2\pi}{\log \frac{b}{a}} \leq \iint_\Omega \rho(z)^2 \, dx \, dy$$

for every admissible metric ρ. Taking the infimum over all admissible metrics, we conclude that $\lambda(\Gamma) \leq \frac{1}{2\pi} \log \frac{b}{a}$. To prove equality, we need to display an extremal metric. But the metric

$$\rho(re^{i\theta}) = \frac{1}{\log \frac{b}{a}} \frac{1}{r}$$

is admissible for Γ, and a simple calculation shows that

$$\iint_\Omega \rho(z)^2 \, dx \, dy = \frac{2\pi}{\log \frac{b}{a}}.$$

Thus, for the family Γ of arcs connecting the two circular boundary components of the annulus Ω, the extremal length is $\lambda(\Gamma) = \frac{1}{2\pi} \log \frac{b}{a}$.

Next let Γ be the family of closed curves *separating* the two boundary components of Ω. Since Γ contains in particular each of the circles $\{z = re^{i\theta} : 0 \leq \theta \leq 2\pi\}$ for $a < r < b$, each admissible metric ρ must satisfy the inequality

$$1 \leq \left\{ \int_0^{2\pi} \rho(re^{i\theta}) r \, d\theta \right\}^2 \leq 2\pi r \int_0^{2\pi} \rho(re^{i\theta})^2 \, r \, d\theta.$$

Integration with respect to r now gives

$$\frac{1}{2\pi} \log \frac{b}{a} \leq \iint_\Omega \rho(z)^2 \, dx \, dy,$$

so that $\lambda(\Gamma) \leq \frac{2\pi}{\log \frac{b}{a}}$. To prove equality, we need only display the extremal metric $\rho(re^{i\theta}) = \frac{1}{2\pi r}$, which is admissible for Γ. Thus the extremal length of the family Γ of separating curves is $\lambda(\Gamma) = 2\pi/(\log \frac{b}{a})$. Note that the extremal lengths of the two families of curves in the annulus are reciprocals of each other.

Now suppose that Ω is an arbitrary ring domain, a doubly connected domain whose complement (with respect to the Riemann sphere) does not have a degenerate (or single-point) component. Such a domain Ω can be mapped conformally onto an annulus $a < |w| < b$ with $0 < a < b < \infty$, and the ratio b/a is known to be a conformal invariant determined by Ω. Indeed, the invariance of b/a follows from the preceding discussion of extremal length, since the extremal length of the family of curves joining the two boundary components of Ω is preserved under conformal mapping, and the extremal length of the corresponding family Γ for the annulus $a < |w| < b$ is $\lambda(\Gamma) = \frac{1}{2\pi} \log \frac{b}{a}$.

It is customary to define the *module* of a ring domain Ω as $\mu(\Omega) = \frac{1}{2\pi} \log \frac{b}{a}$ if Ω can be mapped conformally onto an annulus $a < |w| < b$. Thus, the module of a ring domain can be computed directly as the extremal length of the family of curves joining the two boundary components. This formulation leads to good estimates as demonstrated, for instance, in Section 6.2.

Fuchs [1] gives a proof of the Koebe one-quarter theorem by extremal length. He also gives a connection between extremal length and harmonic measure. Ahlfors [2] displays a formula that calculates extremal length through the Dirichlet integral of a certain harmonic measure. The formula allows the capacity of a set in the plane to be expressed in terms of extremal length. Extremal length also plays an important role in the theory of quasi-conformal mappings, especially in its generalizations to higher dimensions. As these remarks suggest, extremal length has a wide variety of applications in complex analysis.

References

Y. Abu-Muhanna and A. Lyzzaik

[1] A geometric criterion for decomposition and multivalence, *Math. Proc. Cambridge Philos. Soc.* **103** (1988), 487–495.

[2] The boundary behaviour of harmonic univalent maps, *Pacific J. Math.* **141** (1990), 1–20.

Y. Abu-Muhanna and G. Schober

[1] Harmonic mappings onto convex domains, *Canad. J. Math.* **39** (1987), 1489–1530.

L. V. Ahlfors

[1] *Lectures on Quasiconformal Mappings* (Van Nostrand, Princeton, N.J., 1966).

[2] *Conformal Invariants: Topics in Geometric Function Theory* (McGraw-Hill, New York, 1973).

[3] *Complex Analysis* (Third Edition, McGraw-Hill, New York, 1979).

Y. Avci and E. Złotkiewicz

[1] On harmonic univalent functions, *Ann. Univ. Mariae Curie-Skłodowska* **44** (1990), 1–7.

J. L. M. Barbosa and A. G. Colares

[1] *Minimal Surfaces in* $\mathbb{R}^3$, Lecture Notes in Math. No. 1195 (Springer-Verlag, Berlin, 1986).

L. Bers

[1] Isolated singularities of minimal surfaces, *Ann. of Math.* **53** (1951), 364–386.

[2] Univalent solutions of linear elliptic systems, *Comm. Pure Appl. Math.* **6** (1953), 513–526.

B. V. Bojarski

[1] Homeomorphic solutions of a Beltrami system, *Dokl. Akad. Nauk SSSR* **102** (1955), 661–664 (Russian).

[2] On solutions of a linear elliptic system of differential equations in the plane, *Dokl. Akad. Nauk SSSR* **102** (1955), 871–874 (Russian).

B. V. Bojarski and T. Iwaniec

[1] Quasiconformal mappings and non-linear elliptic equations in two variables I, II, *Bull. Polish Acad. Sci. Math.* **22** (1974), 473–478, 479–484.

D. Bshouty, N. Hengartner, and W. Hengartner

[1] A constructive method for starlike harmonic mappings, *Numer. Math.* **54** (1988), 167–178.

D. Bshouty and W. Hengartner

[1] Boundary correspondence of univalent harmonic mappings from the unit disc onto a Jordan domain, in *Approximation by Solutions of Partial Differential Equations, Quadrature Formulas, and Related Topics*, B. Fuglede *et al.*, eds., NATO ASI Series C, Vol. 365 (Kluwer Academic Publishers, Dordrecht-Boston-London, 1992), pp. 51–60.

[2] Univalent solutions of the Dirichlet problem for ring domains, *Complex Variables Theory Appl.* **21** (1993), 159–169.

[3] Univalent harmonic mappings in the plane, *Ann. Univ. Mariae Curie-Skłodowska* **48** (1994), 12–42.

[4] Boundary values versus dilatations of harmonic mappings, *J. Analyse Math.* **72** (1997), 141–164.

[5] Exterior univalent harmonic mappings with finite Blaschke dilatations, *Canad. J. Math.* **51** (1999), 470–487.

D. Bshouty, W. Hengartner, and O. Hossian

[1] Harmonic typically real mappings, *Math. Proc. Cambridge Philos. Soc.* **119** (1996), 673–680.

D. Bshouty, W. Hengartner, and T. Suez

[1] The exact bound on the number of zeros of harmonic polynomials, *J. Analyse Math.* **67** (1995), 207–218.

H. Chen, P. M. Gauthier, and W. Hengartner

[1] Bloch constants for planar harmonic mappings, *Proc. Amer. Math. Soc.* **128** (2000), 3231–3240.

G. Choquet

[1] Sur un type de transformation analytique généralisant la représentation conforme et définie au moyen de fonctions harmoniques, *Bull. Sci. Math.* **69** (1945), 156–165.

[2] Sur les homéomorphies harmoniques d'un disque *D* sur *D*, *Complex Variables Theory Appl.* **24** (1993), 47–48.

M. Chuaqui, P. Duren, and B. Osgood

[1] The Schwarzian derivative for harmonic mappings, *J. Analyse Math.* **91** (2003), 329–351.

J. A. Cima and A. E. Livingston

[1] Integral smoothness properties of some harmonic mappings, *Complex Variables Theory Appl.* **11** (1989), 95–110.

[2] Nonbasic harmonic maps onto convex wedges, *Colloq. Math.* **66** (1993), 9–22.

J. Clunie and T. Sheil-Small

[1] Harmonic univalent functions, *Ann. Acad. Sci. Fenn. Ser. A.I* **9** (1984), 3–25.

F. Colonna

[1] The Bloch constant of bounded harmonic mappings, *Indiana Univ. Math. J.* **38** (1989), 829–840.

H. L. de Vries

[1] A remark concerning a lemma of Heinz on harmonic mappings, *J. Math. Mech.* **11** (1962), 469–471.

[2] Über Koeffizientenprobleme bei Eilinien und über die Heinzsche Konstante, *Math. Z.* **112** (1969), 101–106.

U. Dierkes, S. Hildebrandt, A. Küster, and O. Wohlrab
[1] *Minimal Surfaces I* (Springer-Verlag, Berlin-Heidelberg-New York, 1991).

M. J. Dorff
[1] Some harmonic n-slit mappings, *Proc. Amer. Math. Soc.* **126** (1998), 569–576.
[2] Convolutions of planar harmonic convex mappings, *Complex Variables Theory Appl.* **45** (2001), 263–271.
[3] Minimal graphs in $\mathbb{R}^3$ over convex domains, *Proc. Amer. Math. Soc.* **132** (2004), 491–498.

M. J. Dorff and T. J. Suffridge
[1] The inner mapping radius of harmonic mappings of the unit disk, *Complex Variables Theory Appl.* **33** (1997), 97–103.

M. Dorff and J. Szynal
[1] Harmonic shears of elliptic integrals, *Rocky Mountain J. Math.*, to appear.

K. Driver and P. Duren
[1] Harmonic shears of regular polygons by hypergeometric functions, *J. Math. Anal. Appl.* **239** (1999), 72–84.

P. L. Duren
[1] *Theory of H^p Spaces* (Academic Press, New York, 1970; reprinted with supplement by Dover Publications, Mineola, N.Y., 2000).
[2] *Univalent Functions* (Springer-Verlag, New York, 1983).
[3] A survey of harmonic mappings in the plane, in *Texas Tech University, Mathematics Series, Visiting Scholars' Lectures* 1990–1992, vol. 18 (1992), pp. 1–15.

P. Duren and W. Hengartner
[1] A decomposition theorem for planar harmonic mappings, *Proc. Amer. Math. Soc.* **124** (1996), 1191–1195.
[2] Harmonic mappings of multiply connected domains, *Pacific J. Math.* **180** (1997), 201–220.

P. Duren, W. Hengartner, and R. S. Laugesen
[1] The argument principle for harmonic functions, *Amer. Math. Monthly* **103** (1996), 411–415.

P. Duren and D. Khavinson
[1] Boundary correspondence and dilatation of harmonic mappings, *Complex Variables Theory Appl.* **33** (1997), 105–111.

P. Duren and G. Schober
[1] A variational method for harmonic mappings onto convex regions, *Complex Variables Theory Appl.* **9** (1987), 153–168.
[2] Linear extremal problems for harmonic mappings of the disk, *Proc. Amer. Math. Soc.* **106** (1989), 967–973.

P. Duren and W. R. Thygerson
[1] Harmonic mappings related to Scherk's saddle-tower minimal surfaces, *Rocky Mountain J. Math.* **30** (2000), 555–564.

R. Finn and R. Osserman
[1] On the Gauss curvature of non-parametric minimal surfaces, *J. Analyse Math.* **12** (1964), 351–364.

C. FitzGerald and Ch. Pommerenke
[1] The de Branges theorem on univalent functions, *Trans. Amer. Math. Soc.* **290** (1985), 683–690.

W. H. J. Fuchs
[1] *Topics in the Theory of Functions of One Complex Variable* (Van Nostrand, Princeton, N.J., 1967).

S. Gleason and T. H. Wolff
[1] Lewy's harmonic gradient maps in higher dimensions, *Comm. Partial Differential Equations* **16** (1991), 1925–1968.

G. M. Goluzin
[1] *Geometric Theory of Functions of a Complex Variable* (Moscow, 1952; German transl., Deutscher Verlag, Berlin, 1957; Second Edition, Izdat. "Nauka", Moscow, 1966; English transl., American Mathematical Society, Providence, R.I., 1969).

M. R. Goodloe
[1] Hadamard products of convex harmonic mappings, *Complex Variables Theory Appl.* **47** (2002), 81–92.

A. W. Goodman and E. B. Saff
[1] On univalent functions convex in one direction, *Proc. Amer. Math. Soc.* **73** (1979), 183–187.

P. Greiner
[1] Boundary properties of planar harmonic mappings, Ph.D. Thesis, University of Michigan, 1995.
[2] Geometric properties of harmonic shears, *Computational Methods and Function Theory*, to appear.

A. Grigoryan and M. Nowak
[1] Integral means of harmonic mappings, *Ann. Univ. Mariae Curie-Skłodowska* **52** (1998), 25–34.
[2] Estimates of integral means of harmonic mappings, *Complex Variables Theory Appl.* **42** (2000), 151–161.

A. Grigoryan and W. Szapiel
[1] Two-slit harmonic mappings, *Ann. Univ. Mariae Curie-Skłodowska* **49** (1995), 59–84.

R. R. Hall
[1] On a conjecture of Shapiro about trigonometric series, *J. London Math. Soc.* **25** (1982), 407–415.
[2] On an inequality of E. Heinz, *J. Analyse Math.* **42** (1982/83), 185–198.
[3] A class of isoperimetric inequalities, *J. Analyse Math.* **45** (1985), 169–180.
[4] The Gaussian curvature of minimal surfaces and Heinz' constant, *J. Reine Angew. Math.* **502** (1998), 19–28.

W. K. Hayman
[1] *Multivalent Functions* (Cambridge University Press, London, 1958).

E. Heinz
[1] Über die Lösungen der Minimalflächengleichung, *Nachr. Akad. Wiss. Göttingen Math.-Phys. Kl.* (1952), 51–56.
[2] On one-to-one harmonic mappings, *Pacific J. Math.* **9** (1959), 101–105.

W. Hengartner and G. Schober
[1] On schlicht mappings to domains convex in one direction, *Comm. Math. Helv.* **45** (1970), 303–314.
[2] A remark on level curves for domains convex in one direction, *Appl. Analysis* **3** (1973), 101–106.
[3] Univalent harmonic mappings onto parallel slit domains, *Michigan Math. J.* **32** (1985), 131–134.
[4] On the boundary behavior of orientation-preserving harmonic mappings, *Complex Variables Theory Appl.* **5** (1986), 197–208.
[5] Harmonic mappings with given dilatation, *J. London Math. Soc.* **33** (1986), 473–483.
[6] Univalent harmonic functions, *Trans. Amer. Math. Soc.* **299** (1987), 1–31.
[7] Curvature estimates for some minimal surfaces, in *Complex Analysis: Articles Dedicated to Albert Pfluger on the Occasion of his 80th Birthday*, J. Hersch and A. Huber, editors (Birkhäuser Verlag, Basel, 1988), pp. 87–100.
[8] Univalent harmonic exterior and ring mappings, *J. Math. Anal. Appl.* **156** (1991), 154–171.

W. Hengartner and J. Szynal
[1] Univalent harmonic ring mappings vanishing on the interior boundary, *Canad. J. Math.* **44** (1992), 308–323.

J. Hersch and A. Pfluger
[1] Généralisation du lemme de Schwarz et du principe de la mesure harmonique pour les fonctions pseudo-analytiques, *C. R. Acad. Sci. Paris* **234** (1952), 43–45.

S. Hildebrandt and F. Sauvigny
[1] Embeddedness and uniqueness of minimal surfaces solving a partially free boundary value problem, *J. Reine Angew. Math.* **422** (1991), 69–89.

D. Hoffman
[1] The computer-aided discovery of new embedded minimal surfaces, *Math. Intelligencer* **9** (1987), 8–21.

E. Hopf
[1] On an inequality for minimal surfaces $z = z(x, y)$, *J. Rational Mech. Anal.* **2** (1953), 519–522; 801–802.

J. M. Jahangiri, C. Morgan, and T. J. Suffridge
[1] Construction of close-to-convex harmonic polynomials, *Complex Variables Theory Appl.* **45** (2001), 319–326.

S. H. Jun
[1] Harmonic mappings and applications to minimal surfaces, Ph.D. Thesis, Indiana University, 1989.
[2] Curvature estimates for minimal surfaces, *Proc. Amer. Math. Soc.* **114** (1992), 527–533.
[3] Univalent harmonic mappings on $\Delta = \{z \; : \; |z| > 1\}$, *Proc. Amer. Math. Soc.* **119** (1993), 109–114.

[4] Planar harmonic mappings and curvature estimates, *J. Korean Math. Soc.* **32** (1995), 803–814.

D. Khavinson and G. Swiatek

[1] On the number of zeros of certain harmonic polynomials, *Proc. Amer. Math. Soc.* **131** (2003), 409–414.

H. Kneser

[1] Lösung der Aufgabe 41, *Jahresber. Deutsch. Math.-Verein.* **35** (1926), 123–124.

J. Krzyż and M. Nowak

[1] Harmonic automorphisms of the unit disk, *J. Comput. Appl. Math.* **105** (1999), 337–346.

R. S. Laugesen

[1] Injectivity can fail for higher-dimensional harmonic extensions, *Complex Variables Theory Appl.* **28** (1996), 357–369.

[2] Planar harmonic maps with inner and Blaschke dilatations, *J. London Math. Soc.* **56** (1997), 37–48.

O. Lehto

[1] *Univalent Functions and Teichmüller Spaces* (Springer-Verlag, New York, 1987).

O. Lehto and K. I. Virtanen

[1] *Quasiconformal Mappings in the Plane* (Second Edition, Springer-Verlag, Berlin-Heidelberg-New York, 1973).

H. Lewy

[1] On the non-vanishing of the Jacobian in certain one-to-one mappings, *Bull. Amer. Math. Soc.* **42** (1936), 689–692.

[2] On the non-vanishing of the jacobian of a homeomorphism by harmonic gradients, *Ann. of Math.* **88** (1968), 518–529.

H. Liu and G. Liao

[1] A note on harmonic maps, *Appl. Math. Lett.* **9** (1996), no. 4, 95–97.

A. E. Livingston

[1] Univalent harmonic mappings, *Ann. Polon. Math.* **57** (1992), 57–70.

[2] Univalent harmonic mappings II, *Ann. Polon. Math.* **67** (1997), 131–145.

A. Lyzzaik

[1] On the valence of some classes of harmonic maps, *Math. Proc. Cambridge Philos. Soc.* **110** (1991), 313–325.

[2] Local properties of light harmonic mappings, *Canad. J. Math.* **44** (1992), 135–153.

[3] The geometry of some classes of folding polynomials, *Complex Variables Theory Appl.* **20** (1992), 145–155.

[4] Univalence criteria for harmonic mappings in multiply-connected domains, *J. London Math. Soc.* **58** (1998), 163–171.

[5] A note on the valency of harmonic maps, *J. Math. Anal. Appl.* **218** (1998), 611–620.

[6] The modulus of the image annuli under univalent harmonic mappings and a conjecture of J. C. C. Nitsche, *J. London Math. Soc.* **64** (2001), 369–384.

O. Martio

[1] On harmonic quasiconformal mappings, *Ann. Acad. Sci. Fenn. Ser. A.I*, (1968), 3–10.

M. Mateljević and M. Pavlović
[1] Multipliers of H^p and $BMOA$, *Pacific J. Math.* **146** (1990), 71–84.

A. D. Melas
[1] An example of a harmonic map between Euclidean balls, *Proc. Amer. Math. Soc.* **117** (1993), 857–859.

Z. Nehari
[1] The Schwarzian derivative and schlicht functions, *Bull. Amer. Math. Soc.* **55** (1949), 545–551.
[2] *Conformal Mapping* (McGraw-Hill, New York, 1952; reprinted by Dover Publications, New York, 1975).

G. Neumann
[1] Valence of complex-valued planar harmonic functions, *to appear*.

J. C. C. Nitsche
[1] Über eine mit der Minimalflächengleichung zusammenhängende analytische Funktion und den Bernsteinschen Satz, *Archiv der Math. (Basel)* **7** (1956), 417–419.
[2] On harmonic mappings, *Proc. Amer. Math. Soc.* **9** (1958), 268–271.
[3] On an estimate for the curvature of minimal surfaces $z = z(x, y)$, *J. Math. Mech.* **7** (1958), 767–769.
[4] On the constant of E. Heinz, *Rend. Circ. Mat. Palermo* **8** (1959), 178–181.
[5] On the module of doubly-connected regions under harmonic mappings, *Amer. Math. Monthly* **69** (1962), 781–782.
[6] Zum Heinzschen Lemma über harmonische Abbildungen, *Arch. Math. (Basel)* **14** (1963), 407–410.
[7] *Lectures on Minimal Surfaces*, Vol. I (Cambridge University Press, Cambridge, 1989).

M. Nowak
[1] Integral means of univalent harmonic maps, *Ann. Univ. Mariae Curie-Skłodowska* **50** (1996), 155–162.

R. Osserman
[1] On the Gauss curvature of minimal surfaces, *Trans. Amer. Math. Soc.* **96** (1960), 115–128.
[2] Global properties of classical minimal surfaces, *Duke Math. J.* **32** (1965), 565–573.
[3] *A Survey of Minimal Surfaces* (Second Edition, Dover Publications, Mineola, N.Y., 1986).
[4] Minimal surfaces in R^3, in *Global Differential Geometry*, S. S. Chern, editor, Mathematical Association of America Studies in Mathematics Vol. 27 (1989), pp. 73–98.

G. Pólya and G. Szegő
[1] *Isoperimetric Inequalities in Mathematical Physics* (Princeton University Press, Princeton, N.J., 1951).

Ch. Pommerenke
[1] *Univalent Functions* (Vandenhoeck & Ruprecht, Göttingen, 1975).

T. Radó
[1] Aufgabe 41, *Jahresber. Deutsch. Math.-Verein.* **35** (1926), 49.

[2] Über den analytischen Charakter der Minimalflächen, *Math. Z.* **24** (1926), 321–327.

[3] Zu einem Satze von S. Bernstein über Minimalflächen im Grossen, *Math. Z.* **26** (1927), 559–565.

E. Reich

[1] The composition of harmonic mappings, *Ann. Acad. Sci. Fenn. Ser. A.I* **12** (1987), 47–53.

[2] Local decomposition of harmonic mappings, *Complex Variables Theory Appl.* **9** (1987), 263–269.

H. Renelt

[1] *Quasikonforme Abbildungen und elliptische Systeme erster Ordnung in der Ebene* (B. G. Teubner, Leipzig, 1982); English edition: *Elliptic Systems and Quasiconformal Mappings* (John Wiley & Sons, New York, 1988).

M. S. Robertson

[1] Analytic functions star-like in one direction, *Amer. J. Math.* **58** (1936), 465–472.

W. C. Royster and M. Ziegler

[1] Univalent functions convex in one direction, *Publ. Math. Debrecen* **23** (1976), 339–345.

W. Rudin

[1] *Principles of Mathematical Analysis* (Third Edition, McGraw-Hill, New York, 1976).

St. Ruscheweyh and L. Salinas

[1] On the preservation of direction-convexity and the Goodman-Saff conjecture, *Ann. Acad. Sci. Fenn. Ser. A.I* **14** (1989), 63–73.

L. E. Schaubroeck

[1] Subordination of planar harmonic functions, *Complex Variables Theory Appl.* **41** (2000), 163–178.

[2] Growth, distortion and coefficient bounds for plane harmonic mappings convex in one direction, *Rocky Mountain J. Math.* **31** (2001), 625–639.

G. Schober

[1] Planar harmonic mappings, in *Computational Methods and Function Theory*, Lecture Notes in Math. No. 1435 (Springer-Verlag, Berlin-Heidelberg, 1990), pp. 171–176.

J. T. Schwartz

[1] *Nonlinear Functional Analysis* (Gordon and Breach, New York, 1969).

H. S. Shapiro

[1] Research problems in complex analysis (edited by J. M. Anderson, K. F. Barth, and D. A. Brannan), *Bull. London Math. Soc.* **9** (1977), 129–162; Problem No. 7.26, p. 146.

T. Sheil-Small

[1] On the Fourier series of a finitely described convex curve and a conjecture of H. S. Shapiro, *Math. Proc. Cambridge Philos. Soc.* **98** (1985), 513–527.

[2] On the Fourier series of a step function, *Michigan Math. J.* **36** (1989), 459–475.

[3] Constants for planar harmonic mappings, *J. London Math. Soc.* **42** (1990), 237–248.

J. J. Stoker

[1] *Differential Geometry* (Wiley-Interscience, New York, 1969).

D. J. Struik

[1] *Lectures on Classical Differential Geometry* (Second Edition, Addison-Wesley, Cambridge, Mass., 1961; reprinted by Dover Publications, Mineola, N.Y., 1988).

T. J. Suffridge

[1] Harmonic univalent polynomials, *Complex Variables Theory Appl.* **35** (1998), 93–107.

T. J. Suffridge and J. W. Thompson

[1] Local behavior of harmonic mappings, *Complex Variables Theory Appl.* **41** (2000), 63–80.

A. Szulkin

[1] An example concerning the topological character of the zero-set of a harmonic function, *Math. Scand.* **43** (1978), 60–62.

J. L. Ullman and C. J. Titus

[1] An integral inequality with applications to harmonic mappings, *Michigan Math. J.* **10** (1963), 181–192.

N. Vekua

[1] *Generalized Analytic Functions* (Pergamon Press, London, 1962).

A. Weitsman

[1] Harmonic mappings whose dilatations are singular inner functions, preprint (1996).

[2] On the dilatation of univalent planar harmonic mappings, *Proc. Amer. Math. Soc.* **126** (1998), 447–452.

[3] On the Fourier coefficients of homeomorphisms of the circle, *Math. Res. Lett.* **5** (1998), 383–390.

[4] On univalent harmonic mappings and minimal surfaces, *Pacific J. Math.* **192** (2000), 191–200.

[5] Univalent harmonic mappings of annuli and a conjecture of J. C. C. Nitsche, *Israel J. Math.* **124** (2001), 327–331.

[6] On the Poisson integral of step functions and minimal surfaces, *Canad. Math. Bull.* **45** (2002), 154–160.

W. L. Wendland

[1] *Elliptic Systems in the Plane* (Pitman, London, 1979).

A. Wilmshurst

[1] The valence of harmonic polynomials, *Proc. Amer. Math. Soc.* **126** (1998), 2077–2081.

J. C. Wood

[1] Lewy's theorem fails in higher dimensions, *Math. Scand.* **69** (1991), 166.

Index